# When The Bugle Calls

## A Soldier's Memoir

**Ray Boland**

Cover design by Hemingway Publishers

ISBN: Printed in the United States

Library of Congress Registration Code: NRC112952

# Dedication

To Jackie and Mike:

They fought like brave soldiers in battles against cancer that could not be won. Their final bugle calls sounded much too soon.

*"The greatest danger for most of us is not that our aim is too high and we miss it, but that it is too low and we reach it."*

*- Machiavelli*

# CONTENTS

Dedication iii

About the Author vii

Introduction viii

Chapter 1 The Old Chicago Neighborhood 2

Chapter 2 The United States Enters World War II 13

Chapter 3 Life as a Harper Cardinal 31

Chapter 4 On Wisconsin 61

Chapter 5 Back to College 86

Chapter 6 The Berlin Crisis 117

Chapter 7 Fort Rucker Flight School 138

Chapter 8 Hanau, Germany 149

Chapter 9 Good Morning, Vietnam 180

Chapter 10 Home From War 204

Chapter 11 The Vietnam Screaming Eagles 220

Chapter 12 The Cav 242

Chapter 13 Career Soldiers 260

Chapter 14 Tropic Lighting 276

Chapter 15 A Future in the Army 298

Chapter 16 Rock of the Marne 320

Chapter 17 A Dangerous World 339

Chapter 18 The Last Parade 355

Epilogue 385

Acknowledgments 390

Acronyms 392

Photo Gallery 398

# About the Author

Raymond G. Boland is a 30-year Army veteran with two tours of duty in Vietnam. While there, he flew U1-A Otter utility airplanes to remote Special Forces camps. Ray also flew AH-1 Cobra attack helicopters as a unit commander in the 101st Division.

Born in Chicago, Ill., in 1937, Ray moved with his family to Friendship, Wis., following his high school graduation in 1955. He became a public school teacher and taught at several rural central Wisconsin schools. Ray enlisted as a private in the Wisconsin Army National Guard in 1956 and was called to active-duty in 1961 during the Berlin Crisis. He remained on active-duty until his retirement as a Colonel in 1991. Ray earned a Master of Science Degree in Communications from Shippensburg State University, Pa., and is a graduate of the Army War College. His military awards include the Legion of Merit, the Distinguished Flying Cross, The Bronze Star, the Purple Heart, and the Meritorious Service Medal.

Following military service, Ray completed a second career in the Wisconsin State Government as the Secretary of Veterans Affairs and retired in 2003. He resides with his wife, Donna, in Sparta, Wis.

# Introduction

I was just four years old when Japan bombed the United States military bases at Pearl Harbor on December 7, 1941. This day is one of those days that will forever be etched in my memory. Even though I was not old enough to read or write, I clearly remember how frightened my mother was. At this young, vulnerable age, I was not sure what I should be afraid of. Yet, Mom's sense of angst quickly became a badge of fear that was very real in my tiny body. I do not believe this transfer of alarm was intentional. It just happened.

As I look back to those early days after the Japanese attack, I see certain seeds planted within me. At this young age, I did not know how to respond. Or what this attack truly meant for the United States (U.S.) and the rest of the world. Nor did I have any idea how one single event would have so much influence on me. Now, I see how those seeds planted so early in my life carried me toward a very specific and certain direction for the next 50 years.

Nearly 20 years later, on January 20, 1961, the newly elected President of the United States, John F. Kennedy, delivered his inaugural address in Washington D.C. During this speech, his words issued a profound challenge to all Americans when he said, "Ask not what your country can do for you. Ask what you can do for your country." Earlier in this same speech, President Kennedy made a call, saying, "Now the trumpet summons us again." I was 23 years old at the time. The words came to my heart like a trumpet. They beckoned me to answer the call. As this trumpet spoke to my heart, I knew I was being called to serve my country. The only question was whether I would answer the trumpet's call.

The call of the trumpet is something that has deep biblical roots and many meanings. When the bible was written, trumpets were often used to communicate a message. Throughout the Old Testament, trumpets announced impending war, a coming of joy, the onset of peace, a warning of destruction, and much more to the Jewish nation.

I find an interesting parallel between the symbolism of biblical trumpets and the modern-day playing of the military bugle. Military bugle calls are heard throughout the day as they announce different messages that prompt action. During the Civil War, bugle calls were used extensively to send messages and commands on the battlefield.

A familiar bugle call today is the playing of *Taps* at night and *Reveille* in the morning. These calls signal the end of one day and the beginning of another. They may be viewed as God's message that each day will end, regardless of how difficult it has been. We are promised the sun will rise in the morning and encouraged to seek hope for a new day.

These simple truths can be a powerful source of strength when soldiers are in the darkest moments of combat. I witnessed this firsthand and drew upon these same certainties for strength as an active-duty soldier in the U.S. Army from 1961 to 1991. My years of service included the entire Cold War period and two years in Vietnam as an Army aviator.

This book is my story of answering the bugle's call and how it shaped me as a person. This summons influenced my life long before I enlisted and became an official member of the U.S. Army. It is written as accurately as I remember it, with uncensored candor and brutal honesty. The events found within these pages are true, although some stories may appear stranger than fiction.

This story is not just about me. It is written to honor all the women and men who have served in our armed forces, past and present, as well as the many hardships their families faced. It is a tribute to the United States of America for its incredible and unwavering commitment towards world freedom and peace, as today's young people still answer the bugle call to serve their country on foreign soil. This is their story, too. And yes, it needs to be shared.

Every soldier has a story to tell. Unfortunately, most are never told. For many years, my family urged me to write this story and share it with others. While serving as a soldier has been my greatest honor, this path has been riddled with experiences that have saddened my heart and mind. Like many soldiers, I was reluctant to talk about my military experiences. It felt easier to not share them than try to explain the situations. This is partially why this project took over ten years to complete.

One of my favorite quotes comes from Indira Gandhi, who says, "Nothing that is worthwhile is ever easy." There are parts of this story that are anything but easy. Putting these words into a format that captured the spirit of how they happened and their profound effect on my life was more challenging than I anticipated. Yet, I encourage other veterans to tell their stories. Without these efforts, generations of important experiences will perish, and mounds of reflection and understanding will be lost.

During my active-duty service with the Army, my three sons grew up moving from one Army post to the next. They experienced a lot of my story firsthand. Constant uncertainty is stressful and downright difficult. We experienced happy times, sad times, funny times, and lonely times with extended family separations. Today, we talk mostly about the good times and how unique the whole experience was for us. During the tough times, we tried to maintain our sense of humor. The ability to poke fun at ourselves and with each other was a constant in our lives. This became our way of dealing with the reality that I was called to serve our country as a soldier. Our home was where the Army sent us. Our family was the people we served with.

My call for you as you read these words is to reflect upon the various challenges you have faced in your life and how the bugle may have summoned you. Life can take many pathways. My journey included military service. As I share how the bugle summoned me, allow it to beckon you to learn, explore, and encounter your own story. Reach inside. Recall where you started and how you determined where you were going. What did it take to get you there? Are you still traveling the road? What did you hope to find when you arrived? What lessons did you learn along the way? Writing this book has helped me reflect upon the bugle calls in my life. I invite you to learn, explore, and encounter your own story as I share some of mine.

# Chapter 1

# The Old Chicago Neighborhood

How much does where you grow up influence your childhood? I believe a whole bunch! At least, this was my experience. Reflecting upon my childhood from the vantage point of a rearview mirror, I have come to appreciate how the neighborhood where I was raised planted the seeds that would eventually become part of the bugle's call in my life.

I was raised in the West Englewood neighborhood on the south side of Chicago, Ill. It was a simple and ordinary place. Growing up there in the 1940s started me on an improbable life journey that I could not have planned or even dreamed of. If I yearned to go back and start over, I doubt there is much I would change. Now, it seems like it was all meant to be.

By today's standards, my childhood was remarkably simple. We did not have television in our home. After school, my brother and I listened to radio programs like *The Lone Ranger*. There wasn't any jet aircraft. Space travel was a science fiction subject about a guy named Buck Rodgers. We had no computers, hand-held calculators, or even electric typewriters. Social media consisted of a wall-mounted telephone connected to a switchboard operator who would say, "Number, please." If you were on a party line shared with neighbors, you might hear them talking when you picked up the receiver. Sunday family get-togethers at Grandma's house were a big deal because they usually included a five-cent ice cream cone from the little store across the street. Lots of things I liked cost five cents back then. A bottle of Pepsi, a Baby Ruth candy bar, a pack of baseball cards, bubble gum, and a Popsicle. All the true pleasures in life for a seven-year-old cost just one nickel. The exception was a White Castle Hamburger, which cost a whopping dime.

While I could go on about the simplicity of life in the old days, there are a few specific events that influenced my journey and the trajectory it took. A big one was World War II. It would be hard for today's generation to imagine how much the war period affected everyone's life at that time. It was a very frightening time in my life, but also the one that inspired me to serve my country later. Nothing I have seen since compares to the fears our country felt.

Other influences included my parents, Ray Sr., Ann, and the church I attended. My parents were members of the Greatest Generation. They survived the depression era by working ridiculously hard at any job they could find. Neither of them graduated from high school, although my father valued formal education. An average-looking guy from his era, Dad had a slightly longer face and a strong forehead. He also held strong religious beliefs. My mother was taller than my dad and used her smile to her advantage. She was fiercely competitive and expected me to be the best in everything. My father only expected me to try my best at whatever I did. These contrasting and competing influences were something I struggled to balance. Fortunately, my church helped to put things in perspective and allowed me to begin to understand who I was.

For the first 17 years of my life, I lived on or near South Damen Avenue in Chicago. The city is laid out with all the north-south streets having names and those running east-west being numbered. I was born near 51st, and Damen and last resided in Chicago at 47th and Damen. My farthest south childhood home was near 69th Street. The family command post was my grandparents' house at 6717 South Damen. Almost everything I did happened within a stretch of 20 blocks, including where we lived, went to school, did our shopping, and attended church. My world was exceedingly small. In many respects, this may not have been much different than growing up in a small town or even a rural area. I assume that most youngsters during my childhood grew up within limited geographic boundaries.

My outside world consisted of the area between South Ashland Avenue and South Western Avenue, which ran parallel to Damen. This is where my family went for serious shopping. Both of those streets were lined with places of business. Damen was mostly a lazy residential street with only a few stores. But 69th Street, near Damen, was a mecca of small mom-and-pop shops, including a bakery, butcher shop, hardware store, drug store, and barber shop. An Italian enclave along 69th Street included two grocery stores and the then-famous Louis George restaurant. Ambrosino's and Casio's sold freshly made Italian sausage, with cheeses hanging from above the counters and sensational aromas that tickled your nose as soon as you opened the door. A Chicago fixture at the time was the neighborhood tavern. Damen did not have many of those. "Ozzie" Oswald's tavern and beer garden at the corner of 69th and Damen was an exception. There were no strangers in this neighborhood. Just like on the television show *Cheers*, everyone seemed to know each other's names.

I did not realize how limited my lifestyle and knowledge of the world around me was because I was happy. Nothing else mattered much. With no electronic media other than radio and no smartphones, our communications mainly were interpersonal and word of mouth. We had no choice but to talk to each other. This way, we were shaped by very few influences.

I recently visited the Empire State Building in New York City for the first time in more than 60 years. Everything seemed pretty much as I remembered it, with one stark exception. As the crowd walked through the maze towards the final elevator, I noticed that nearly everyone in line was staring into a hand-held device. Hardly anyone spoke to another person. For the most part, they seemed unaware that other people were there. Everywhere I go today, I see people texting. It seems very odd compared to my childhood.

In those early years of my existence, the strongest influence in my life was my mother. I do not think this was unique to my upbringing, as many children from my generation had a similar experience. She was an incredibly unique person. According to her mother, she was the leader of a pack in her younger days. Grandma called them the "Damen Devils." I never did quite figure out who the "Devils" were, but I got the drift. My mother claimed she was the toughest kid on the block. She prided herself on being able to outdo any boy in everything. Out run, out smart, out challenge, outfight, outlast. From time to time, I overheard her siblings recall stories of some of her escapades, which actually verified her claims.

My mother was like no other person I had known or heard of. It is difficult to underestimate how I have been swayed by the lessons I observed from her. She instilled her strengths and weaknesses in me when I was young. It took a long time for me to understand which were which. My favorite cousin still lives in the Chicago area and remembers experiencing the "wrath of Auntie Ann."

A complete analysis of her colorful personality is not necessary. Yet, the way she influenced me is what I think is important in this story. Some of it was unusual, with both positive and negative outcomes. These elements stayed with me for a long time. Today, like other older folks, I have mellowed a lot, as did my mother in her final years. By then, I knew that I loved her very much. When Mom died in 2000, my respect and admiration for her as a parent and as a person was at its peak. With maturity, I realized any unpleasant memories I had from the past were about things she did, believing they were for my own good.

My father was raised by his Irish grandmother near 51st and Damen. They were extremely poor. He shared stories of nearly being caught stealing coal along a nearby railroad track to heat their basement apartment. Growing up, he saw little of his mother. He never met his father. Thus, he had no real family until he married my mother. His grandmother required him to regularly attend a Lutheran church and become a Boy Scout. He enjoyed both very much, and those influences shaped him tremendously. In time, these same organizations had a significant impact on me. Everyone liked my father. He was kind, considerate, talented, and a lot of fun. I admired his fairness, compassion, and good nature.

Dad connected with my mother through two of her brothers. One was a former junior high school classmate, and the other brother attended the same church. The brothers invited Dad to their house, where he met their sister. Immediately, he began his new life. My parents were married in 1936. Their first home was a small upstairs apartment at 51$^{st}$ and Damen. I was born there in 1937.

## Our Family

I would like to claim I can recall everything that happened from the beginning. I must confess my mother kept a detailed diary from the time I was born. It has wonderfully enhanced my memory. Two significant things happened when I was only two years old. The first was a serious vehicle accident in December 1939. It almost killed my father. He was working for the U.S. Post Office as a mail truck driver in downtown Chicago when he was struck by another vehicle. His vehicle was turned over. Dad was pinned under the truck and sustained severe internal injuries. It was an extremely tough time for my mother, who was pregnant with my brother. After a lengthy coma and hospitalization, he fully recovered. It was considered a miracle that he pulled through this ordeal. In a photo taken on Easter 1940, he looks very weak. My brother Alan was born in June 1940.

Mother's diary says that, like many other two-and-a-half-year-olds, I was not exactly thrilled to have the competition of a new baby in the house. She also explains that eventually, my brother Al and I grew together and enjoyed each other a lot. I am happy to say that the same feeling continued. Al and I always felt close despite being apart geographically for most of our lives.

After his recovery, Dad returned to work as a postal employee and eventually received a government retirement pension. In 1941, we moved into a rented brick bungalow at 68th and Damen. This was a big step up. There were few brick buildings along Damen, as most were aged wooden structures.

In our new home, we were one block away from Grandma's house. Only a half block away was the Marlboro Presbyterian Church, where my father enrolled me in Sunday School when I was five. I continued in this small neighborhood church until my late teens. It became an important refuge from the other aspects of my world. There was a sense of safety and security there that made me feel protected from the pressures I felt at home. As a teenager, I became active with youth fellowship group activities throughout the Chicago area, including retreats at the Druce Lake Camp.

These were years when temptations fueled by peer pressure could easily take one in the wrong direction, just as they can today. Sharing a spiritual environment with kids of my age kept me on track. We were blessed to have an energetic and youth-oriented pastor who spent a lot of time with us. We also had a couple of incredible group leaders who were excellent role models. At the time, I did not realize how significant and permanent the value of my church experience would be. When I look at children in church today, I see them having the same experience. This is an important dynamic that has not been altered by time, technology, and the increasingly fast pace of life. It is something that can benefit every child, regardless of their economic status.

## Common People

At an early age, I heard about families being described as upper class, middle class, or lower class. I remember asking my mother what we were. She said we were middle class. Today, I think the working class would be a fair label for our status and many of those around us. The four words I remember most from my childhood are, "We can't afford it." There were no credit cards in those days nor limited time payment options. If you did not have the money to buy something, you lived without it. So, we did not have much. In his entire lifetime, my father never owned a new car. After he returned from World War II, he always had an evening part-time job in addition to his full-time job as a mail carrier. He would leave early in the morning before my brother, and I awoke. He came home at night after we were in bed. On weekends, he worked at a gas station and tended bar. There were extended periods when my brother and I saw very little of him.

Today, I occasionally meet someone who lives in a larger area of Chicago. When I explain the location of my roots, their typical reaction is, "Wow, you wouldn't want to live there now."

I usually reply, "You probably wouldn't have wanted to live there then because it wasn't that great."

Several years ago, my current wife and I drove to Chicago. I gave her the grand tour of Damen Avenue and each of the places where I lived. One of the houses we resided in, along with my grandparents' house, was vacant and boarded up. Many other buildings were uninhabited, and others were torn down. None of the little stores at $69^{th}$ and Damen were still there. The few remaining buildings along $69^{th}$ Street appeared uninhabited. At one point in the tour, my wife said she was glad we had not heard any gunfire. The next morning, we read that a fatal shooting occurred the night before, one block away from where we drove. Today, the West Englewood district of Chicago, my old neighborhood, is one of the most violent crime areas in the country.

The old neighborhood I knew is gone forever. There are few of us left who remember what it was like. It was a vastly different time in the history of our country. Most of our grandparents came from the Old Country and tended to reside among ethnic neighborhood clusters such as Italian, Irish, Polish, Lithuanian and others. It also contributed to the scope of our upbringing. We were all impressed with a strong work ethic. The value of hard work took priority over social status.

There is a saying that you can never go back to things the way they were. I certainly cannot because the South Damen I knew no longer exists. I do believe that growing up there as part of the working class fueled my life's journey. It gave me tools that outlasted the demise of the neighborhood. The values and ethics we learned from common people in a very ordinary neighborhood took us to wherever our future level of effort aspired. For me, it eventually took me towards the bugler's call. However, other influences entered my life first.

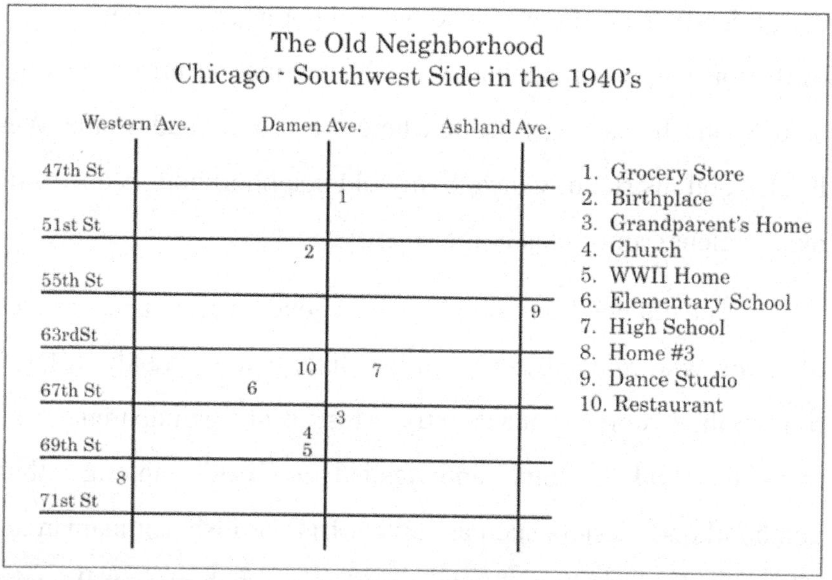

# Chapter 2

# The United States Enters World War II

Combat is not the only place where people experience the repercussions of war. The effects of war happen around the kitchen table where grocery lists are written and based upon what remains on ration cards. It happened within backyard or front porch conversations conveying details provided by letters received from loved ones serving in Europe or the Pacific Theatre. We relied upon reports heard on the radio, where families gathered around and listened carefully to weekly updates.

My initial understanding of the United States military began when Pearl Harbor was bombed. On that Sunday, I remember a sense of panic setting in, and fear of what might follow ran wild in my mind. I asked my mother why something that happened so far away was so dangerous to us. As the following days and months passed, we learned more about why this was such a catastrophic event.

The war dominated every aspect of American life for the next four years; few other things mattered. A spirit of patriotism abounded, and Americans became enormously proud of every man and woman who wore military uniforms. Like so many of my playmates, I acquired my first soldier suit and wore it often, including to school. Every school day began with reciting the *Pledge of Allegiance* and singing '*My Country Tis of Thee.*' I do not recall hearing any objections to this from classmates, teachers, or parents.

By first grade, I had learned how to salute and respect the American flag and sing all the patriotic songs. A favorite of mine was the Army Air Corps song. I was fascinated with military airplanes and constantly doodled pictures of them. I clearly remember when my mother told me I would become a combat pilot someday. I did not anticipate how accurate her prophecy would be!

In third grade, our teacher talked about what it meant to be a good citizen. She expected every student to behave with no questions asked. At the end of the semester, she awarded the American flag displayed at the front of the classroom to the student who modeled the best citizenship. I was immensely proud when I marched home with the flag. Loyalty to my country would never decrease from this point forward throughout my life.

My father had a very gung-ho military attitude. He liked wearing uniforms and was very proud of his mail carrier uniform and badge. He taught me to be respectful of all police officers and firefighters. After World War II started, Dad became highly active with the Civil Defense Corps and served as a block captain to educate and coach our neighbors on preparing for air raids. There was significant fear that either Germany or Japan might bomb us. Maybe both. Air raid drills were practiced regularly at home and school.

Quickly, the casualties of war also hit home. Gold stars soon appeared in home windows, indicating the family had lost someone in the war. Our family was worried that my father would be drafted. Initially, his being drafted seemed unlikely because he was 26 and married with two children, and the draft focused on younger, single men.

Two years later, when the United States mobilized forces for the invasion of Europe, it became a much different situation. My father was drafted. When my mother learned this news, she became frantic. Mother insisted that he should be eligible for a deferment. Even if he could evade the draft, I doubt he would have. Like many Americans at this time, my dad desperately wanted to serve our country in a time of greatest need. He and so many others felt this was their duty. We thought he would not be called up as a U.S. Postal employee. Ironically, this ultimately became the reason he was.

The European invasion force included a huge logistics tail. Mail from home was a significant morale booster for the troops.

New Army postal units were formed by drafting experienced mail carriers. Before deployment, these draftees only needed basic combat training and a crash course on the Army postal system.

I clearly remember the last night he was home before he was shipped out. He reported to Fort Sheridan, Ill., on Christmas Day 1943. On Christmas Eve, my brother and I were crying and distraught. Dad sat between us, and he had his arms around us, trying to console us. My father promised us he would be back when the war ended.

I remember saying, "Daddy, what will happen if we lose the war?" His reply has stayed with me until today. He said, "Boys, we are the United States of America, and we will not lose the war." I was only six years old, but I had already memorized the words for the U.S. national anthem, *The Star-Spangled Banner*. I heartily sang the words with gusto and enthusiasm but with little understanding of what the words meant. On this particular night, I began to realize and understand exactly what the statements "the land of the free" and "the home of the brave" meant. My daddy was leaving to defend our country and to make sure our country remained free. To me, this was the definition of bravery. Thoughts of Christmas presents diminished because I was more focused on how proud I was of my dad.

Once my dad left, our family was in a significant financial

bind. My mom, brother, and I were now definitely qualified as poor. My mother went to work full-time as a waitress. My brother and I spent a lot of time with our grandmother. I was expected to become a big boy quickly, just like so many other kids. I started attending kindergarten before my fifth birthday. My mother began teaching me to read at home earlier than this. I started second grade when I was just six years old and could already read quite well. Because of all the time I spent with my grandparents, I also learned to speak and read the Czech language. Even though they had immigrated to the U.S. around the turn of the century, they spoke no English. My grandparents became U.S. citizens in 1921.

Originally from the land known as Bohemia, my grandparents referred to themselves as Bohemians. They were clear that it was not acceptable to identify their ethic group with the Slovakians or, God forbid, the Polish! They genuinely believed the Bohemians were the elite people group from this part of the world.

Coincidentally, I started hearing about politics, racial and religious prejudice after the war began. I doubt today's advocates for social justice can appreciate how intense these issues were in the 1940s. Racial prejudice was widespread and openly discussed at home. There is no question we were taught the use of the N-word and used it daily. The most positive reference I heard in reference to Black people was when they were called "Colored." Chicago public schools were highly segregated, which left no opportunity to become acquainted with any Black kids and become friends. It just did not happen. Thus, ignorance prevailed.

With ethnic neighborhood enclaves of Italians, Greeks, Slovaks, Lithuanians, and others throughout the city, ethnic slurs were common but considered less offensive than racial ones. Some of these groups fought to preserve their all-white neighborhoods.

For a reason I never learned, my mother's family denounced Catholicism. Although they were not regular church attendees, my mother was baptized Catholic. Most curious to me now is that they became staunch Republicans. At the time, there was constant rhetoric about the Democrats sticking up for the working class and the Republicans being only for the rich. I clearly remember being privy to these conversations some 80 years ago. None of our family group seemed to have anything against rich folks. They also wanted to become rich by living out the American dream. My family distrusted Democrat politicians because they believed Democrats did not care about the poor. My family felt Democrats said what was necessary to secure people's votes.

Grandpa Dusek worked at a major factory as a drill press operator. In the 1940s, his plant became unionized. He was approaching retirement eligibility. Grandpa refused to join the union. There was a lot of family discussion about this decision. Grandpa's sons became worried for his safety as he received threats from union leaders at the plant. The family urged him to "just go along" with the times and ride it out until he retired. Grandpa stubbornly refused to consider joining the union. He could not understand or accept why he should forfeit any of his hard-earned money to a group that had done absolutely nothing to make him the outstanding employee he was.

Grandpa retired in 1945. He worked at the plant for over 40 years and never joined the union. In retirement, Grandpa perfected a vegetable garden in a vacant lot beside his house. Everyone who walked by it paused to admire it. Despite an unhealthy diet, smoking, and almost no exercise, Grandpa lived to 84.

Still a youngster in single digits throughout the war, the ongoing political discussion was confusing to me. As the war drew down, the rhetoric increased. Disappointed that President Roosevelt and his successor, President Truman, sold out the Eastern European countries to Russia, my family lost respect and no longer trusted them. There was much sorrow and tears when all the relatives in the old country became locked behind the Iron Curtain. My elderly family members found it unbelievable that the U.S. would turn these countries over to another dictator after all the sacrifices the U.S. made to achieve victory and free people in Europe. Honestly, I do not think the mature members of my family ever recovered from the disappointment they felt.

## Victory in Europe

When my mother passed away in 2000, my family came across a lifetime of memorabilia she kept. In addition to a diary about me, we found all my report cards from elementary school. She kept many letters, postcards, and photographs my dad sent from overseas. In his letters, Dad carefully censored what he shared and kept his comments more general. Dad also kept a diary of his war experiences, which I find remarkably interesting. Unlike his letters, his thoughts are uncensored in his diary. He describes locations and events in great detail, including what he saw culturally and the attitudes of local people.

When I read *Killing Patton,* written by Bill O'Reilly, I was fascinated with the similarity of my father's sharing of events he was aware of with the description recorded in this book. Dad's Army Postal Unit supported the Ninth Army. His diary includes a chronology of events and movements on the ground precisely as described in O'Reilly's book. The Battle of the Bulge was a major turning point in stopping the final German offensive. After this battle, the war headed toward its conclusion.

As the war wound down, Dad's postal unit continued to follow behind combat units until the war ended. Dad wrote about the people he met in Holland, Belgium, France, and Germany. Repeatedly, he shared how much they appreciated American soldiers' presence in their countries.

This similar reaction has been shared and repeated several times with other wars in which the U.S. has been involved. These comments were remarkably similar to what I gleaned from a World War I veteran friend. They also compared my observations in Vietnam, as well as reports I have heard from those who served in Iraq and Afghanistan.

I believe the American GI has been the world's most excellent ambassador for more than one hundred years. Everywhere their boots have tread, GIs have left a footprint of freedom. They represent a pathway that cannot be rubbed out and have provided hope to many millions of people for a better way of life. When people in oppressed nations have seen Americans in person and witnessed what they represent firsthand, non-Americans are determined to no longer accept the oppression they have lived with for years. This long reach of history supports this conclusion, while sometimes, the short-sighted sound-bite mentality of daily media does not.

Ray Fuller was a World War I Veteran from Wisconsin. We became friends in his later years. He lived until he was 100 years old and was a great inspiration to me. Fuller eloquently spoke of this topic during a ceremony at the Wisconsin High Ground Memorial when the World War 1 statue was dedicated. He said, "Did I and hundreds of thousands of other good, young Americans do what we did just to make the world safe for Standard Oil and those others who make up the Fortune 500? And should this statue be dedicated today in order to remind those who look at it that anything and everything is worth sacrificing for American business success? I would not agree."

He continued, "What was my war about? The names and dates of battles, the weapons used, and the medals awarded are the least of them. My war should be remembered as the first time that part of the world learned what America and Americans were really like. As soldiers, we set an example, not just for worthy achievements in war, but for what all Americans have been, are now, and can be. America is a nation of people willing to accomplish what they've been asked to do, no matter what the odds against us."

I believe soldiers, from the First World War doughboys to the soldiers who served in Afghanistan, represent a century of incredible, selfless sacrifice borne by the men and women who served in our armed forces. I strongly believe their legacy as ambassadors of freedom is unmatched throughout history. It is a permanent legacy of the American soldier's dedication to living out their mandate to serve the land of the free and the home of the brave.

## The Farm

When school let out for the summer of 1944, my brother and I were separated. The adults thought the two of us were too much for Grandma to manage. I was literally farmed out for the summer to a Wisconsin farm. At the ripe age of six, I boarded a train headed to northern Wisconsin with Grandma. We traveled to her cousin's farm near Phillips, Wis. After a few days, she returned to Chicago. I found myself completely alone with strangers on a primitive farm. There was no running water, and they relied solely on a wood stove for cooking. It is correct to assume that this was a memorable experience.

Homesick may not be the accurate word to describe my time away. My family did not have much home life in Chicago. I was lonely on the farm and missed my brother Alan very much. We played together quite well, and it was awfully hard to be apart. He was only four, and the adults thought the farm would be too much for him. They were right.

My saving grace while at the farm was a teenage daughter who became my big sister. Ellie was wonderful. I admired her very much. She possessed some of the fiery Damen Devil spirit. She rode a horse and shot a rifle better than any boy. At 15, she drove the relic farm car with no brakes. She milked the cows, pitched hay, and did it all. Most days, I followed her around, trying to help her out by milking cows, pitching hay, gathering eggs, and carrying water. The next summer, Alan traveled with me to this same farm. Again, Ellie was the highlight of the summer. After the second year on the farm, I only saw her one more time before she passed away.

A few years ago, I visited this farm site. All that remained was the land. The buildings were in rough shape when I last saw them in 1945. I was not surprised they were all gone. I have some photos from when I stayed there. I consider the outhouse the most nostalgic building on the farm.

While the farm buildings were non-existent, the old Midway Tavern across the road was still standing. The sight of this building drew me back to those late Saturday nights decades ago when I lay in bed, wide awake, listening to the thump of the bass drum from the polka band playing at a wedding dance at the tavern. Such events are only memories today.

## Dancing With the Stars

As a teen, my mother had an obsession to be in show business. I never understood this. She ran away from home once with a touring dance troupe but returned home as a failure. It was a humiliating experience for her. I am still convinced her biggest wish was that I would be a girl and live out her fantasy. Instead, I showed up as a baby boy. Yet, Mom stayed determined to have a family star in show business.

I was just five years old when she enrolled me in tap dancing lessons. As a little shaver, I was kind of cute and did my best to please her. Over time, I no longer thought dancing was charming and certainly could think of other things I preferred to do with my time. Nonetheless, I danced for another ten years. Every Saturday, I took private lessons from a former vaudeville hoofer. He was a nice guy, but I could tell dancing was not my gig. It did not help that I felt dancing was a waste of my time. I did not see how my skills would improve enough to achieve my mother's dream for me.

At 14, I knew of only two other boys on the entire south side of Chicago who were tap dancers. One of them told me he was taking lessons from a terrific Black dancer in downtown Chicago. Fed up with dancing, I knew my only chance of advancing was to train at a higher level. My request to train with this man was immediately dismissed because of his skin color. My mother's ambitions for me to become a top-notch dancer would not include training by a Black instructor.

Without an instructor of a higher caliber, I knew my chances of becoming an exclusive tap dancer were impossible. I would never become like the elite skilled male dancers of the day: Ray Bolger, Fred Astaire, and Gene Kelly. Some experts rated Buddy Ebsen as the best dancer of all. His notoriety came not from dancing but as Jed Clampett in *The Beverly Hillbillies*.

With an alternative teacher no longer an option, I informed my mother that I was quitting dancing. She became terribly upset and made me feel guilty for wasting all the hard-earned money she had invested in my lessons. The conversation was not pretty. After she calmed down, Mom asked me to wait a bit longer while she talked over the whole situation with my instructor.

The following week, I went back to my instructor. He had an alternative idea for me to consider. There was a new female student who was exceptionally talented. He thought we would make a great dancing duo. He asked me to come back the following week and meet her. Reluctantly, I agreed.

Rita was a beautiful girl and so very pleasant. It was all too perfect. She was someone it would be hard to resist getting to know. Immediately, I sensed the meaning of psychological warfare. This effort was simply to impress upon me my mother's dream. The teacher asked Rita and me to consider trying a few routines together. The following week, I went back and announced my decision. I was done with tap dancing. Forever. It was an early test of having the willpower to do what I thought was best. I broke my mother's heart. It would not be the last time this would happen.

Years later, I ran into Rita while waiting for a train at a Chicago L station. She looked even prettier than I remembered and still seemed extremely sweet. We exchanged small talk for a few minutes and wished each other well. I never saw her again.

## Post-War Transition

After the war, my dad came home. While our family was together again, I could sense that things had become strained between my parents. Only later in life, when I experienced this myself, did I understand why. Extended separation is unhealthy for partners. This is one of the heavy tolls of military service. In recent years, multiple deployments of our troops to Iraq and Afghanistan have stretched many relationships to the breaking point. We seldom hear of the cost of active-duty couples' relationships. Personally, I am aware of countless marriages that did not endure the Vietnam era.

My dad planned to return to his post office job, and he did. He appreciated the security and benefits of a government job. My mother had vastly different ideas. She believed the only way they could get ahead was to own a business. She resented that my dad was not more ambitious. As time progressed, this sticking point became a serious disagreement, which led to more challenging times for our entire family. My dreams of our family being happily together once Dad was home were quickly dashed.

In retribution to Dad's decision to return to his previous position, Mom opened a small restaurant at 66th and Damen. A restaurant is a generous description of her business. It was a lunch counter in a poor location with no parking and no realistic chance of succeeding.

In reality, the restaurant was a front for a horse racing bookie operation Mom also ran. While I assisted her with the analysis of the performance records of different horses, I was not fully aware of what was going on behind the scenes. My involvement only included analyzing the horses. Using the information published in the *Racing Form*, we tracked horses in great detail. Each evening, we updated our records. Our objective was to know when and under what track conditions a horse achieved their best times for a particular distance. The daily scratch sheets I retrieved from a local newspaper stand were also essential. These helped us know which horses were at what odds for a race each day.

I found this all remarkably interesting, but I still did not know what was really going on. My involvement was limited, and I didn't have a full understanding of how well-orchestrated my mother's business had become. Restaurant customers stopped by for a bowl of chili or a piece of pie and coffee. While eating lunch or dessert, they placed their bets. Many other bets were placed by phone. Pies were delivered daily by the pie man in a rack of stacked pie tins with the betting money placed between the tins. Outgoing bets were under empty pie tins. Winning monies were placed under fresh pie tins. It was a very clever arrangement and the one that I suspect was used all over Chicago.

When my dad realized what was going on, he knew it was only a matter of time before Mom and her affiliates would get caught doing something very illegal. He knew that this had to stop. I do not know what the tipping point was, but suddenly, the restaurant closed. We moved to a family home near 69th Street. This was the first house my parents owned.

I was relieved with this move. I convinced myself that the idea of owning a business was out of Mom's system. Unfortunately, this was not the case. There was much more to come. My mother's distinctive personality and desire to make her mark would continue to influence my childhood.

## A Changed World

As the 1940s decade transitioned toward the 1950s, the U.S. and the world had undergone major changes. The U.S. had grown into an industrial giant and world superpower. So many things in everyday life had changed. When the automobile industry resumed production in 1946, cars looked exactly the same as in 1942. By 1949, a drastic rebirth of the industry happened. The days of the original Henry Ford design no longer existed. The Rocket 88 Oldsmobile engine made the flat-head V-8 obsolete. With the birth of overhead valves and high-compression engines, the age of muscle cars was born. Televisions were now in the majority of homes. We watched the Chicago White Sox play without going to Comiskey

Park. Milton Berle's *Texaco Star Theater* show was a family favorite.

While it seemed that post-World War II Americans yearned for a time of peace and internal focus, the reality was quite different. The nuclear age had begun. With it came the fear of another potential nuclear war. Communism replaced fascism as the global threat to peace. For the second time in the 20$^{th}$ century, we had suffered through another "war to end all wars." This was supposed to be it. Never again would we need a strong military. Or so most Americans thought. Time and history told a different story. The bugle's call would soon send another generation of soldiers to war.

# Chapter 3

# Life as a Harper Cardinal

"We're from Harper; couldn't be prouder. If you can't hear us, we'll yell a little louder!"

When I graduated from McKay Elementary School in January 1951, I was barely 13 years old. Chicago public schools operated on a semester system. I wound up starting high school much too early. I was only four when I started kindergarten. In fourth grade, I jumped ahead a semester with an early promotion into fifth grade. By the time I entered the pearly halls of Harper High School at 6520 South Wood Street, I was a whole school year ahead of my age group.

Harper was one of the smallest public high schools in Chicago, with an enrollment of 1,300. Nonetheless, it felt huge and overwhelming to me. With a trimester schedule, June classes usually had about 250 students. A January freshman class had only about one hundred students. I was the youngest and shortest kid in my class. I was still a soprano and a chubby tap dancer who withstood a lot of teasing from classmates; it was not an enjoyable time for me. High school years can be difficult for anyone. Growing up is rarely easy. Some of my ups and downs were not much different from the experiences of other kids. Most teenagers tend to believe their situation is more complicated than their peers. My early entry into high school and immaturity were a distinct liability. For the remainder of my high school years, I was rarely a very happy camper.

## Joining the Junior ROTC

I enrolled in the high school's Junior ROTC class as an alternative to gym class. Because of my mother's strong personality, she made most of the family decisions. Enrolling me in Junior ROTC was one of the few decisions my father had a major hand in. With my limited physical makeup, sports were not on my horizon despite my rising interest. My dad really liked being a soldier and wanted to stay in the Army. When this did not happen, he dreamt of me following in his footsteps. Thus, he became actively involved in making sure I joined the Junior ROTC.

Before my first day of high school and Junior ROTC class, Dad put me through a basic at-home training program. He followed the drill and ceremonies training by the book. I used a broom handle to learn the manual of arms. He taught me how to execute and recite the commands. I knew all movements to standard before I walked into my first ROTC class. Already, I was an officer waiting to be commissioned. My dad was extremely proud of me and my preparations.

Initially, I excelled in ROTC and did well in my other studies. However, my accelerated school progress and immaturity caught up with me. When I became eligible for promotion to Corporal, I knew my name had to be near the top of the student ROTC list. I boasted this to my fellow cadets. Some of this was overheard by the cadre, and I was not promoted. It was one of the first major blows to my ego. Even worse, I felt that I had been cheated. All my hard work felt unrewarded. My resentment showed, which only pushed me further toward the rear of the ROTC bus.

About the same time, I realized I was not as smart as my mother thought I was. I struggled in class. Always strong in reading, reading comprehension, and social studies, my aptitudes in math and science were soft in grade school, but I kept up. Now, as the difficulty factor quickly accelerated, I knew it would require demanding work for me to keep up. It seemed like too much effort, and I became lazy. Until then, everything in school had been too easy. I was warned about this. Unfortunately, it was one of those messages I did not want to hear, and I ignored it.

I watched a varsity basketball game during my first high school semester. Immediately, I was fascinated. In those seemingly dark ages, there was no public-school basketball before high school. While my grade school had a gym with baskets, we never played basketball. The skills and teamwork I witnessed during this first game were much more intense than what I had seen with baseball.

Rapidly, I was hooked. I was determined to become a basketball player.

I had no idea what would happen, but I convinced myself that I should try. This was the first time I felt a sense of conviction to accomplish something, no matter how difficult it would be. I gave up dancing because I no longer wanted to dance. Basketball became something I felt I wanted to do. Quickly, it became something I had to do.

A few Catholic friends told me about an open gym at the nearby Catholic high school for the kids of our age. One Father was a staunch basketball fan. He opened up the gym and supervised pickup games on Saturday afternoons. My friends knew I was dying to learn basketball. They said they would vouch for me with the father. One Saturday afternoon, I was allowed to go. My mother warned me to be home by 6 p.m.

We had a fantastic time. I ran up and down the court with the other kids. Everyone had so much fun. Time quickly slipped away, and Father was not in a hurry to kick us out. Suddenly, it was 6 p.m. Still perspiring, I ran out of the gym, jumped on my bicycle, and sped home. I arrived about 10 minutes late. I knew I would be in trouble, but I never expected the door to be locked. It was a brisk November evening. I began to shiver. I knocked repeatedly on the door and pleaded with my mother. She ignored me and pretended like she did not know I was on the other side of the door.

After 10 or 15 minutes, I was cold and started to panic. Fortunately, a neighbor lady walked across the street to visit my mother. When she saw me, she asked, "Raymond, why are you out in the cold without warm clothes?"

When I told her that my mother had intentionally locked me out of the house, she was in disbelief. She rang the doorbell and pounded on the door. With no answer, she became irate and started yelling. After she said who was at the door, my mother quickly opened the door. Our neighbor lady went into a tirade and threatened to call the police. I became scared, knowing I now faced a double whammy for all the trouble I had caused. This poor start in my quest to become a basketball player did not deter me from my goal. It made me even more determined to make the team.

This incident caused me to reflect upon a situation that happened a few months earlier during summer vacation. In violation of another curfew, my mother came to a vacant corner lot where I was playing softball to enforce her order. I was at bat and did not see her come up behind me. She was wielding a switch. When she drew the switch to hit me, the other kids yelled, "Watch out!" As I turned around to see what was happening, I saw the switch coming towards me. I held the bat in front of me for protection. The switch struck the bat, broke it in two, and injured my mother's hand. Later, my mother said her hand was permanently damaged and declared it was all my fault.

As basketball season was about to begin in the fall of 1951, I asked the coach if I could try out for the team. Completely unequipped skills-wise, the coach's gracious reaction surprised me. While he appreciated my interest, he told me that I was not ready to try out. Instead, he extended me an alternative offer. His friend coached a team in the recreation league. He would arrange for me to join this park league. If I promised to "bust my butt and do everything I could to learn" from his friend, he would let me try out for the school team the following year. I was elated.

I was still a short, chubby kid who had never played basketball. What I did not know is that I had fast feet and the agility to move and change direction quickly. Finally, I realized a blessing from all those years of dancing. My exceptional foot work enabled me to be a better-than-average defense player, and this skill got me on the floor. I did not score a single point the entire season, but my strong defense skills earned the coach's respect and encouragement. I accomplished the first step toward my goal.

In April 1952, my sister Marylu was born. She rightfully received my parents' full attention. I found relief from some of the negative attention often sent my way. Marylu quickly became the apple of my father's eye. My brother and I also adored her, and this has continued through the present. My dad loved singing to her, so he regularly serenaded her with an old Frank Sinatra song that began, "Mary Lou, Mary Lou, cross my heart, I love you." These words, repeated thousands of times, always seemed so fitting.

Because of the age difference and my early departure from our home nest a few years later, I missed much of Marylu's growing-up years. Her history does not include the early stories shared within these pages. Today, she has a family of her own with three beautiful daughters and eight grandchildren. We are blessed to live just fifty miles apart.

In the summer months after Marylu's birth, I grew eight inches. I spent every available moment on the playground or in the alley by our house, where a basketball hoop was mounted to the garage. My mother despised the fact that I spent so much time doing something so useless, in her opinion. She felt jumping up and down with a ball and pretending someone else was there by making head fakes signified mental illness. More than once, she threatened to call an ambulance and have me put in a strait jacket. She was serious.

I returned to school in the fall as a second-semester sophomore. Some of my classmates did not recognize me. It was exciting to think the new me might have a chance to make the Junior Varsity basketball team. Because I kept my promise to the coach, I was hopeful I would get the opportunity to try out. Coach held up his end of the bargain and gave me a chance. I made the Junior Varsity squad. My still-small-chest felt bigger the day I went home in my very own Harper Cardinal uniform. My mother was quite underwhelmed with my efforts and success. To cement her feelings, she told me I would be responsible for laundering my own athletic apparel.

Another aggravation coincided with my making the basketball team. Pulling the plug on tap dancing had removed the growing problem of ridicule among my guy friends. Now, I found myself on the topic of a new hassle among the school athletic crowd. They thought it was very uncool to be in ROTC while also trying to be a jock. Being called names like "tin soldier" was among the nicer things I heard.

Unfortunately, my stock within the Junior ROTC continued a downward spiral after I was passed over for promotion. Basketball practice conflicted with rifle team practice. Achieving membership on the rifle team was an important steppingstone for advancement within the Junior ROTC. I was not achieving the required benchmarks.

By 15, I realized that I had another life dilemma. I could not remain in ROTC and achieve the basketball goals I had for myself. Quitting ROTC would break my father's heart. Already out of favor with my mother, I did not want another broken-hearted parent.

Eventually, I approached my father for a face-to-face discussion. As usual, he patiently listened and digested my sense of frustration. He seemed to understand some of my feelings. Then, he helped frame things into a fuller perspective. Dad was supportive of my goal to succeed in basketball. But he also thought it would be unwise for me to quit ROTC. He did not make me feel guilty. Instead, he used the situation to mentor me.

Dad said this was an opportunity to become a "real man." I could be successful in both. Quitting one would be an admission that I could not achieve my dreams. Dad cautioned me about falling into a pattern of quitting things when the going gets tough. He did not mention the dancing episode, but the message was loud and clear. Quitting was the easy way out.

Looking back, I see the profoundness of this conversation and the way my dad guided me through this decision. While the choice remained mine, Dad's logic made sense to me. I did not want to be defined as a quitter. Even though I was still a teenager, I longed for my dad's approval and wanted to demonstrate that I could do hard things. I could be a real man. I decided to stay the course with both activities.

During the fall semester of 1953, I reached my goal of being selected for the varsity basketball team. I had more confidence in myself than I ever had since starting high school. With an improved attitude, I held my own in ROTC. The church youth group was a wonderful place for fellowship. Life felt good. It was manageable. I hoped to achieve the goals I had established for myself.

But this new-found rhythm was short-lived. At that same time, my mother informed me that I needed to start working. She wanted me to appreciate what it took to earn a living. It was time for me to get a job after school hours. It would be my responsibility to earn any money I needed to spend.

My mother's new directive was a huge wake-up call. Initially, I completely failed to grasp the full meaning of this bugle call. Everything began to pile up on me. I found it increasingly difficult to keep up with my studies, maintain the ROTC schedule, work, and keep up the necessary basketball skills required to keep my spot on the varsity squad. Suddenly, my hard work to succeed in basketball seemed insignificant. I quit the basketball team.

Without the passion for basketball in my life, I began to feel again that my high school days had reached rock bottom. I no longer cared much about anything. Every day after school and on weekends, I stocked groceries at a supermarket and worked as a cashier at a drug store. I went through the motions in class but was a depressed young man. I regretted quitting basketball and had many second thoughts about how I could have managed things differently. This time, my father did not lecture me on quitting. Like my mother, he believed this was the best decision.

## The Grocery Store

Not deterred by the closing of the restaurant/horse racing booking business, my mother decided to try another business. In the summer of 1953, she bought a small neighborhood grocery store on the corner of 47th and Damen. To finance the purchase, my parents sold the house I was once locked out of. Our family moved into the very cramped living quarters behind the store. Marylu was only one year old. There was very little space for my parents, a baby beginning to walk, and two growing boys.

Mom had chosen another business location that had no parking and little potential for success. It should not be a surprise that exactly one year later, she decided to try another business adventure. This time, she wanted to move to Wisconsin and build a motel. By now, I seriously started to think about how I might move away from my parents.

During the summer of 1954, we took a brief family trip to Wisconsin. Mom found a property she liked, and my parents bought it. They planned to sell the grocery store and move to Wisconsin in the summer of 1955. Immediately, I felt great angst about my parents' decision. I was not on board moving to Wisconsin.

## The National Guard's Calling

It was an early spring day in April 1954. While leaving school for work, I saw some military equipment in the schoolyard. A couple of uniformed guys were nearby. Curiosity got the best of me, and I stopped to see what was going on. The guys were from a local National Guard unit from the south side of Chicago. They brought the equipment to my school as part of a recruiting trip.

I was in awe of their static display, which included a 105 mm towed howitzer and a two-and-a-half-ton truck prime mover. The battery first sergeant was in charge of the visit. He gave me a quick sales pitch on the value of joining the National Guard versus an Army Reserve unit. He stressed a significant difference. National Guard units can only be called to active-duty with the activation of the entire unit, which, he said, would never happen. Army Reserve members could be called up individually, which happened regularly. He felt the best way to avoid the current draft and potential active-duty service would be to join a National Guard unit.

Ultimately, his pitch was more about draft evasion than a call to help keep the United States as "the land of the free" or urging to be part of the "home of the brave." Nonetheless, my early military breeding had me intrigued. Those impressive pieces of equipment and the thought of wearing a grown-up military uniform versus the smaller version I had worn in grade school captivated my attention.

The battery first sergeant knew that I was only 16. I knew the minimum age for enlistment was 17. Nonetheless, he quickly enlisted me into the unit. When I questioned how I could join at my current age, he told me he would get me a waiver. Later, I learned the waiver was done with what was known as the M-1 pencil. In other words, my initial enlistment documents were falsified. Many years later, I obtained a copy of those records. Sure enough. There was a pencil change to my date of birth.

I knew enlisting at 16 was not right. I should have waited another year and enlisted legally when I was 17. But for years, I had heard so many stories about war heroes who lied about their age to join the service. These young soldiers had made a difference in previous wars. Later, they were admired and recognized for their service to our country despite breaking the rules. I thought it was kind of cool. I was ready to be part of their club.

Another item on the first sergeant's sales pitch was a two-week all-expense paid vacation, or more accurately, annual training. In July, we would travel to a beautiful place in Minnesota named Camp Ripley. He assured me that I would love it there. New recruits would get to do lots of fun things. While there were many things to complete before the camp began, he assured me that he could get me prepared.

This was possible because, in 1954, there was no requirement for formal training for reserve component enlistees in advance of enlisting. All training was done on the job or OJT. Just a year later, there was a major change in this policy. Since I enlisted before this policy change, I attended two-hour drill meetings one night a week at an armory at 51st and South Cottage Grove in preparation for the Annual Training at Camp Ripley. When I attended drill meetings at this armory, I noticed that it still had stables that were used for horse-drawn artillery. New recruits were issued World War II hand-me-down fatigues and a set of used khakis.

I learned a lot in the months before I went to Camp Ripley. I became a licensed military vehicle operator, received live fire individual weapons familiarization training, and learned how to crew a 105 mm howitzer. Already proficient in close-order drills, I quickly became an accepted and appreciated member of a very understrength unit.

This battery was authorized for around 100 personnel. When we went to annual training in July, 47 people from our battery participated. We took just four of our six guns with a skeleton crew for each. I was now an official cannon cocker.

As our convoy rolled away from the south side of Chicago and drove the 625 miles to Camp Ripley, I still anticipated the paid vacation the first sergeant had promised. Instead, the next two weeks were beyond grueling. During the journey, I was assigned as an assistant driver for a three-quarter-ton truck. The staff sergeant in charge of the truck had his own route plan for the march to Minnesota.

I did not know the staff sergeant had a serious drinking problem. He was unable to go all day without a drink. As we crossed the state line from Illinois into Wisconsin, he applied the truck's choke. This made the engine run rough. Soon, he pulled off the road and onto the shoulder. A vehicle behind us pulled over to see what was wrong. The staff sergeant said the engine was running very rough and cutting out. The people in the other vehicle told us to wait for the maintenance truck towards the end of the convoy.

The staff sergeant in my truck had carefully picked the spot to pull over. It was awfully close to another road he wanted to switch over to. As soon as the vehicles behind us had passed, he sped off onto this different road. He justified his actions by telling me that he knew a shortcut that would get us to our destination by the time the rest of the convoy arrived. I told him that it did not seem right to leave the convoy. He assured me that he knew what he was doing.

Immediately, I knew this was a problem. Even at just 16 years old, I became very worried about what could happen. The convoy was scheduled to spend the night at Camp McCoy in

Wisconsin, about half-way to Camp Ripley. By the time we arrived, we were in big trouble. The higher-ups considered issuing a court marshal. I went from worried to scared in a hurry.

I was interrogated separately from the staff sergeant. I told the person questioning me the truth. He believed my version of the story. I was cleared of any wrongdoing. The staff sergeant was taken back to Chicago. I never saw him again. It was a shaky start for my first full day of military duty.

## No Vacation at Camp Ripley

The next day, when we arrived at Camp Ripley, we settled into our squad tents. Our canvas cots were placed on concrete slabs. Each battery had a row of tents. I was assigned to the Charlie Battery. Using Charlie is a holdover from the World War II phonetic alphabet, which began with Able, Baker, Charlie, Dog, etc. Today's military phonetic alphabet has been updated and begins with Alpha, Bravo, Charlie, etc., with Charlie being the only word carried over into the new alphabet. Ironically, the Charlie Battery Commander and Executive Officer were twin brothers. Their last name was Cannon.

The two weeks at Camp Ripley were no fun or games. The promised vacation was days of grueling drills. We spent many days in the field doing intensive live fire missions. Assigned to the number one position on our gun, I was responsible for setting the elevation, pulling the lanyard to fire the howitzer, and operating the breach block for loading and removing the shell casings. An active-duty Army unit advisor who served in Korean combat tutored me on how to become the fastest number one man in the battalion. It was challenging to meet his standards. He told war stories about firing so many combat rounds in succession that the tube would overheat. They were unbelievable but real.

On firing days, we returned to camp, cleaned the tubes, and ate chow. I usually spent the night walking on guard duty. The next morning, I reported to the mess section for kitchen police or KP duty. Sometimes, I went to the field with the guns. This routine continued non-stop for two weeks: guns, guard duty, KP, guns, etc. The work was hard and exhausting. Just as difficult was learning the military acronyms, what words really meant, and the rules of military life.

When I did get to use the cot in my squad tent, we could hear the first sergeant and unit advisor bellowing long into the night. The more they drank, the louder their voices became. They bad-mouthed everything and everybody. They kept referring to some really bad guy named Divardi. I wondered who this was. He sounded like a really mean Italian gangster. Later, a guy told me the leaders were talking about the Division Artillery Headquarters (div arty.) This was my first experience in discovering that often lower-ranked military personnel complain about their higher headquarters.

Outside of our military drills, I found another way to be accepted within the ranks at Camp Ripley. During high school, I became involved in playing 16-inch softball through the summer park league teams. I had become a fairly decent player. The service battery in our battalion at Camp Ripley had an outstanding softball team and reached the finals of the 33rd Infantry Division company-level championship. The team was one player short. Their recruiting efforts lead them to me. I played right field in the championship game, and we won. Eddie Zolna was the pitcher. He and the shortstop played on a team that won many 16-inch slow-pitch Chicago championships in the following years. Zolna is a legend in Chicago softball history and is a member of the National Softball Hall of Fame.

While in the Army, I played against another softball legend named Eddie Feigner. His four-man team, called *The King and his Court,* traveled around the country, taking on local teams. They played fast-pitch softball, and King Eddie was unhittable. A highlight of his show was to strike out batters while blindfolded with an amazing behind-the-back delivery.

When the two weeks at Camp Ripley ended, I was exhausted. But I was proud that I held up without faltering. At the time, the basic pay for a Private E-1 was $78 a month. Everyone in our battalion received two weeks' pay. While the $39 would not be much for all my hard work, this had definitely been a character-building experience.

On the last day of camp, we were required to report to the pay officer. When it was my turn, I sharply brought my heels together, rendered the proper salute, and said, "Sir, Private Boland reporting for pay." The lieutenant complemented me for having an outstanding camp. Then, he counted out $23 on the pay table. I was stunned and said something like, "Is that all?" He explained about the required deductions for income tax, social security, the old soldier's home, and laundry. Again, he thanked me for a job well done, and I saluted, did an about-face, and walked away. This time, my thoughts were that I *really* had been screwed. I doubted whether the character-building had been worth it.

Upon return to Chicago from Camp Ripley, I found out that my dear, loving, Bohemian grandmother had passed away. The funeral was already over, and I felt terrible that I had missed the service. My parents felt it was best not to try and get me home. My Grandmother meant so much to me. I loved her for all she did for me when I was a young child. Most importantly, she taught me the meaning of love. She did this not with her words but with her deeds. Everyone loved her very much because of the way she lived her life. Prior to her death, she expressed to my mother her fear that once she was gone, her big family would bicker and drift apart. She was correct, as this happened, and I felt an additional loss in that grandma was the only person in the world who could put my mother in her place. Grandmother saved me from many scrapes with my mother. My grandmother was my security blanket when my mom would get upset because I did the simple things kids sometimes do. Now, I felt even more alone.

## The Platoon

I entered my final semester of high school in September 1954. The decision to move to Wisconsin weighed heavily on me. Friends that I had passed up earlier were now on the varsity basketball team. I was the only senior in ROTC who did not advance beyond second lieutenant. I felt disillusioned after the National Guard summer camp. I knew I was adrift. Two positives that kept me going were my church group, and a female classmate I fell in love with over the summer. I could not adequately define or explain this kind of love. But I knew it was different from any other love I had previously experienced. We went steady during our last semester of school, although I had a gut feeling it would not last.

As our final months together passed by, our individual future plans began to take shape. She wanted to go to nursing school but could not afford it. I wanted to go to college but lacked the funds. She decided to secure a full-time job and save for school. I considered staying in Chicago and doing the same, but my family had a new plan. With some financial help from my paternal grandmother, my parents decided I should move to Wisconsin before they did. They wanted me to enroll at the state teacher's college in Stevens Point, Wis., known as the Central State College. The second semester started in February. My marginal C+ high school grades were good enough for me to be accepted.

As this decision weighed on me, another change came through my ROTC program. Ultimately, this change became the most influential experience during my high school years. As the final semester of our senior year began, Master Sgt. Mancini became the new ROTC Department Sergeant. From day one, Mancini was different from his predecessors. A proud soldier and leader, he impressed us immediately and said things would change for the better.

Approximately 45 Chicago high schools comprised my ROTC department. My high school had one of the very smallest programs. Every year, it was a big deal for us to march down Michigan Avenue during the Memorial Day Parade. It was equally impressive for us to see several thousand cadets march together.

Annually, a highly coveted city championship competition was held. While the city rifle team championship was incredibly competitive, the drill platoon competition was the main event. Called the Picked Platoon Competition, the event attracted the absolute best cadets from each school to participate. Our school had a horrible record in previous competitions, which meant there was little interest among us to participate.

Mancini changed this. In his opening remarks to the assembled cadets on the first day of school, he laid out a vision for the school year with specific expectations and goals. He said that we would train to win the city Picked Platoon Competition. At the time, it seemed like a ridiculous statement. But there was an air about him that signaled he was profoundly serious. I only wondered how he would be able to accomplish this. Certainly, I would not be involved.

Once again, I was very wrong. In laying out the plan for this bold operation, he said training would begin the next morning. Each chosen cadet would report to school every day at 7 AM, an hour before classes started. This schedule would continue until the district and city competitions were held in November. He stipulated that all seniors would participate without exception. I thought to myself, "Whoa, this is going too far. He can't make me do this." I just wanted to finish the semester and move on with my life.

Mancini granted my request to speak to him later that day. I shared that I did not want to be a member of the Picked Platoon Competition. My goal was to finish the semester and move on. His response included the most important words I have ever heard.

Mancini saw his job as helping us become leaders. He pointed to the second lieutenant bar that I wore. He said this signified me as a leader and refreshed my memory about what being a leader means. He said the most important quality of a leader is to set the proper example for subordinates. Good leaders do not expect them to do anything they are unprepared to do themselves. Part of this program for four years, I was challenged to lead from the front, not the rear. He said the platoon experience will always be about how a person leads and wins.

Mancini's eyes did not wander. He looked straight at me and emphasized that he was as serious as a heart attack about these views on leadership. Exuding confidence in an extremely focused manner, he was the real deal.

Immediately, I knew the man standing before me would have a profound impact on me. And he did. No one, including my father, had ever said these things to me. I had never thought of myself as a leader. At Camp Ripley, I saw myself only as a laborer. Now, I had an opportunity to train underneath someone who would mold and define me as a person. At this moment, I felt the first sense of a bugle calling me to military service. The call was clear. My only response was whether to heed the call or not. I realized learning leadership was the most important value with which I could leave.

The experience of working with Mancini was nothing short of amazing. We were briefed on how the competition was scored and what was included. The *Field Manual 22-5: Drill and Ceremonies* book became our bible. The competition format was built around a very structured sequence of events. Each platoon was required to follow it to the letter. It included marching while carrying the M-1 Rifle, a prescribed manual of arms sequence, a set sequence of other close-order drill-like facing movements, and a detailed, in-ranks inspection. The platoon was formed to standard with a platoon guide and a platoon leader who gave each of the required commands. This was not intended to be a fancy drill, as portrayed by a demonstration team. It was strictly a by-the-book drill.

Training began with mastering the marching requirements. According to *Field Manual 22-5*, the normal marching rate, or "quick time rate," consisted of 30-inch steps at a pace of 120 steps per minute. During the competition, steps were measured and timed. Our goal was to hit this standard on the nose every time. After enough training, it became almost impossible to do otherwise. Early in our training, Mancini tied a 30-inch rope between the ankles of the platoon guide until he consistently hit this distance. To this day, the platoon guide may not be able to walk any other way.

We spoke the 120-step cadence aloud until we heard it in our sleep. We marched in the long school corridors before other students arrived. We followed 120-foot markers taped to the floor. A stopwatch timed every sequence, and this continued until the competition itself.

The Manual of Arms training was another exercise about precision. As we held our rifles in the port arms position and were commanded to switch to the order arms position, only one loud "whack" could be heard. Every right hand had to hit the upper rifle stock grip with every forearm parallel to the ground at exactly the same time. When this happened, it was a thing of beauty.

Mancini was a perfectionist. Absolutely no detail was left out. He convinced us to buy Army-issued shoes, which were brown in those days. We had polishing sessions to ensure every shoe was exactly the same shade of brown and buffed to an identical state of gloss. This was the finishing touch to be completely uniform. We were the only school that wore identical shoes at the competition.

The platoon leader was another key piece of the plan. Our cadets selected the highest-ranking officer in our unit. He was also the sharpest-looking and most competent leader in our platoon. Tall, slender, and erect, he could have been a poster boy for the Marine Corps. He reminds me of the sentries who command the Tomb of the Unknown Soldier at Arlington Cemetery. Our platoon leader had an outstanding command voice. When he barked the commands, they were loud and clear. He memorized the entire command sequence. Any mistake would be a major score deduction. He was very cool under pressure and did not make mistakes. We felt we were ready as we neared the day for the district competition. After our final dress rehearsal, we knew we could win.

We faced a major challenge in our district from the perennial powerhouse of Lindblom High School. They were our closest neighboring school to the north and a much larger school with around 2,500 students. They also had an incredibly good football team. They always won the district platoon championship and had been city champions as well. Yet, this year, we defeated them in the district championship! We were ecstatic! All of the arduous work and everything we had been taught and practiced had worked. This was such an important lesson for us as future soldiers to learn.

The city competition happened two weeks later. The air was filled with excitement, and the top two teams in each district were eligible for the finals. Lindblom High School would also compete because they placed second in the district competition. Tons of fans and parents cheered on their favorites.

Just before we marched onto the drill floor, I felt nervous. But when I looked at my fellow cadets, I knew that I stood with the winning team. We reached this point together and trusted each other; everyone knew what to do. It was a similar feeling to what I later experienced in combat. When it comes down to crunch time, you must rely on the people around you. Nobody else can help.

The moment of destiny came; it was the only time we would share it together. When our name was announced, the platoon leader said, "Forward, march." As we entered into view, I noticed that a cap fell off the head of someone ahead of me. Wisely, he and everyone else ignored it. We never missed a beat. I hoped a dropped hat would not determine our success. I did not know the scoring system considered this a very minor, non-procedural error with no consequence to our score. The important thing for us was to stay with our timing and not flinch.

During the break between the district and city competitions, some heavy vibes rolled down the street from our nearby rival, Lindblom. Word on the street was they planned to avenge their rare loss. We simply had been lucky in their eyes.

After all the platoons finished, the judges huddled to tabulate the scores. It felt like an eternity before the announcer came to the microphone with the results. He announced Lane Technical School won third place, a huge northside campus with over 4,000 students. I gulped and clutched my sweaty hands. The announcer said the second-place winner was Lindblom High School. Before he could say, "Harper," our fans erupted. They knew Lindblom was still the team to beat. We had overcome them once again.

I cannot describe how good it felt to win this competition. It was like winning the Super Bowl; the culmination of an impossible dream. From that day forward, I believed anything could be accomplished by ordinary people with the right plan and extraordinary effort. Following this experience, I have had many opportunities to apply this same concept for success throughout my military career.

## Post High School

The last two months of high school flew by. Suddenly, it was graduation night. I think the only high school event my parents ever attended was my graduation. They never saw me play in a ball game or win anything. I point this out because most parents did not attend kids' activities like they do today. Whatever we did with extra-curricular activities was okay as long as we did not get into trouble. My parents were not interested in my activities; most parents had not participated in school activities. They had little appreciation for their value. Raised during the Depression, my parents' belief structure said the only important thing was getting a job. Everything else was unimportant and a waste of time.

Today, I feel parents have gone too far the other way. This is just my opinion. There is value in letting kids figure out things on their own rather than having parents hovering over them. Often, parents are too critical of student players and their coaches.

After the graduation ceremony, we went backstage and removed our caps and gowns. I finally realized we were saying goodbye. Suddenly, I realized that this stage of my life was over. I had not thought about what this meant. Even though I had just spent four years with my high school classmates, I did not have any contact with them until 50 years later at a class reunion. I saw my girlfriend only once after graduation.

I had one remaining task to complete before leaving Chicago: inform my National Guard unit that I was leaving. I was asked what “leaving” meant, and I explained that I was going to attend college in Wisconsin. The unit said this was fine, but I would need to drill every Monday evening in Chicago. When I said this was impossible, I was told that I had a problem. Every week, I repeated that I was leaving. Again, I was told that I could not just quit. I never considered the possibility of transferring to the Wisconsin National Guard nor realized there was an artillery unit in Stevens Point. On the night of training before I moved to Wisconsin, I said goodbye. They still did not think I was serious.

After I left, it did not take long for the military police (MPs) to show up at the grocery store on Damen Ave. looking for me. It happened several times before my mother told them she knew I had been enlisted unlawfully. She said she was contacting her congressperson. This ended their visits, and I heard nothing further.

Looking back, the first 17 years of my life seemed to take a long time. There were plenty of twists and turns in my journey. Then, suddenly, it was time to move on. I thought I would return to Chicago, but I never lived there again. As I walked away from Sweet Home Chicago, I also did not realize that I was destined to become a soldier.

I retrospectively view my high school years as a time I grew up. Much of who I would be had been formed during the years I lived in Chicago. I started to realize who I was and can fully understand why some kids wander astray during these formative years. I could have easily chosen a different path when I felt down, discouraged, or even defeated. Most people are able to find the inner strength and faith to realize that life beyond high school holds many opportunities. Our longer road of destiny requires important choices and decisions, especially when the bugle calls.

# Chapter 4

# On Wisconsin

One week after graduation, I boarded a train headed for Stevens Point, Wis. It was a blustery, cold January day in 1955. The temperature high for the day did not get above zero degrees. It was a premonition of many Wisconsin wintery days ahead. Enrolled at the University of Wisconsin-Stevens Point, I found the dormitory had no midyear vacancies. My first order of business was to find a room to rent near the university. The campus housing office provided me with a list of potential rooms. I found a room less than a block away.

It was a private home that rented four upstairs rooms to students. The price was right at $4.50 per week. The catch? The owners not only wanted me to share a room with another student, but I also had to share a bed. Options were extremely limited, but I knew this would not work. The student who rented the room during the first semester was not there. Rumor had it he was quitting school, but he returned briefly just to get his things and dropped out of school. I felt relief when I did not have to share a bed.

Once again, I felt like a complete stranger in a new place. I was utterly lost. I knew no one in Stevens Point. I was younger and more immature than the other students in my classes. Getting physically around campus and the city was a struggle. Feelings reminiscent of how I reacted to the first summer on my relative's Wisconsin farm welled up inside of me. Going to college mid-year, just a couple of weeks after completing high school, and moving to Wisconsin all by myself were not good choices. I was too young and immature and wandered the campus without a compass, not knowing where or how to get anywhere.

Yet, many other kids start college with similar circumstances. I believe so many high school graduates go to college because they are unsure what to do. Other than the absolute best students who are ready for advanced studies of college coursework, I have supported the option of national service after high school for a long time. Military service is not the only option. Other public service tracks are available to teens/young adults who have no clue about what they want to do with their lives. I am not sure if research confirms this, but it seems to me that the performance of college students who did something else for a couple of years before entering college is higher than those who go directly from high school with little conviction about what they want to study. I strongly believe this has been the case with military veterans.

My lack of maturity, loneliness, and lack of direction were confirmed. I was placed on academic probation at the end of my first semester. Thankfully, I secured a summer job at the end of the semester. Truly, it was an answer to prayer. I met an older student at the church in Stevens Point where I attended. He told me about a YMCA children's camp in northern Wisconsin that was hiring college students for the summer. He was a senior staff member at the camp and had good connections with the camp leadership. A position they were trying to fill was a marksmanship instructor for their 22-caliber rifle range. I had gained experience with the .22 caliber rifle while participating in Junior ROTC in high school, even though I had ducked the rifle team in favor of basketball. This staff person asked me to apply, and it was a relief when I was accepted.

As summer drew near, my family was still in Chicago but preparing for their big move to central Wisconsin to build a motel. They had disposed of the grocery business. My father was talked into quitting his post office job and withdrawing his retirement fund. This was the last thing in the world my dad wanted to do, but my mother's wishes prevailed.

With the funds from Dad's retirement, my parents bought a seven-acre property in Central Wisconsin. Located near Friendship, Wis., the property included 250 feet fronting a major state highway. It was a wonderful location with lots of room for a hotel and a parking lot. Because interstate highways had not yet been built, this was by far the best location they had chosen for a business. The property included an old, small farmhouse with brown tar paper shingle siding and a dilapidated one-car garage. It was not much, but it was the site where my parents planned to live out their American dream. Thirty years later, they sold the same property as an established business. Unfortunately, they went through a lot of challenges between when they bought the acreage and when they sold it.

## Breaking Out the Oars

My parents were pleased when I shared that I had a summer job. The outlook for funding my next college semester was bleak, and I needed to save money. I knew the student who was hired to manage the riding stable at the camp. In June 1955, we headed to the Northwoods with the final destination of Camp Manito-Wish. Situated on a lake, the campground is filled with the tallest pines I have ever seen. Campers and staff quickly memorized the camp theme song, *Break Out the Oars*, which is still used to this day.

Every camper who has attended Camp Manito-Wish is trained in canoe tripping, which means they know how to upright and reboard a capsized canoe. It is why every alumnus is known as a "tripper." Adventures into the wilderness, known as "the trail," are wonderful. They are a draw for why people choose this camp. Campers love doing them.

Summer camp life suited me well. I turned the rifle range into a suitable place for campers to learn about and respect guns.

Quickly, all the campers and staff called me Ray the Rifleman. I worked with boys between the ages of 10 and 12. While I was only 17, they saw me as a grown-up college guy. This was my first experience teaching others and would anchor a lengthy list of teaching opportunities that I have had in my lifetime. The campers responded well to my instructions and methodology. Finding my stride at the campground brought stability and encouragement to my life, which previously had felt very disjointed.

Almost immediately, I loved everything about Camp Manito-Wish: the location, the people, the trust in my abilities, and the camper's response to my directions. Most importantly, I relished this place because it promoted spiritual harmony, fellowship, and mutual respect. There was no peer pressure. The religious component of daily camp life was not overdone, yet it kept everyone mindful of God's blessings in our simple lives.

Many of my favorite camp experiences included gathering with campers and staff around a blazing fire for evening sing-a-longs. Like my father, I enjoyed singing but it had always remained hidden behind the shadows of my mother's insistence that I take tap dancing lessons. At camp, our untrained voices joined together harmoniously and could be heard across the lake as we sang a combination of spiritual and camp songs. We were just a bunch of kids and young adults who felt encouraged by the darkened evening skies and a popping campfire to sing from our hearts. The evenings were beautiful and a comforting way to end the day. A couple of counselors played the ukulele. It became a new diversion for me and to this day, I still enjoy occasional plunking.

Elmer Ott was the camp director, a true gentleman with a soft heart for youth. He became a legend because of the thousands of young people he mentored throughout the many years he served as camp director. To this day, I read testimonials about him in the camp newsletter. To say that he has impacted thousands of people is not a stretch. Repeatedly, people share how Mr. Ott's guidance and leadership impacted their lives long beyond their time at camp. As one of the truly fortunate to have known Mr. Ott, the example of how he lived daily would be one I would reflect upon in years to come. He had the uncanny wisdom of knowing when and how to reach campers and young staff members.

One day, I was surprised when Elmer pulled me aside for a personal chat. After he made a few nice comments about my work at the rifle range, he addressed the real reason for the talk. He asked, "Raymond, why don't you take off your mask?"

I responded, "What mask?"

Elmer described in detail an accurate portrayal of who he thought I really was and contrasted it with the person I was pretending to be. Silently, he meticulously observed and analyzed me and my personality. All along, I thought he barely knew who I was. Instead, he had observed how I tried to portray myself as someone I was not. He said I used this as a way to cover up things that I was not happy about.

The camp director said that I played a game and painted myself as being a Joe Cool guy from Chicago. In doing so, I degraded myself and limited my full potential. Instead, he encouraged me to be myself, and he emphasized that I had nothing to be ashamed of. When he said that God had made me a good person, I felt a lump in my throat. Elmer listed the special qualities God has given me and asked me to actually use them. His final words encouraged me to let myself come from behind the mask. If I did so, I would be happier with myself and discover how to use my God-given gifts and talents.

Initially, I felt very defensive and did not want to accept his assessment. As I became more honest with myself, the more I realized he was right. His exact point was that I did not need to be the cocky punk that I was trying to be. It was time for me to put on some big boy pants, grow up, and decide that I wanted to be the person God created me to be.

Soon after that, Mr. Ott summoned me again. I feared another lecture, but his message this time was quite different. It was now the midpoint of the boys' summer camp session. Soon, a new group of boys would arrive. He wanted to promote me to a cabin counselor position. This was a noticeably big step, and I would be fully responsible for eight youngsters, 24 hours a day, for the duration of each camp. Mr. Ott had full confidence that I could manage this responsibility. With a bit of anxiety, I accepted the challenge. My title was converted from Ray the Rifleman to the leader of the Idlewood cabin. Again, I found myself stretched but discovered I could learn new things.

While I was only at Camp Manito-Wish one summer, the lingering call of the Northwoods remains in my heart through the present. This special message was born during the days I spent at camp. The lessons I learned from those few months at Camp Manito-Wish have lasted me a lifetime. Once I became a parent, I hoped to send my own children to Camp Manito-Wish. Unfortunately, multiple moves as an Army family prevented this from happening. More than 40 years after my summer camp experience, I was thrilled to send four of my grandchildren to Camp Manito-Wish. They returned home, gleefully describing their experience much as I hoped: the experience of a lifetime.

These days, the Northwoods call is a significant part of my daily life. Upon my final retirement, I built a home on a lake in the Northwoods. The whistling of the wind in the tall pine trees still sounds as good as ever. I never tire of the tall trees and the hum of a breeze through those trees. When a day ends with a crackling fire, it is always a good day.

## National Guard: Round 2

In September 1955, the United States Congress passed an important legislation known as the Armed Forces Reserve Act. For the first time in U.S. history, all eligible males were required to be available for a six-year military obligation; they would be subject to the draft. Enlistment in reserve component units required the completion of six months of active-duty training as a prerequisite before being deployed to combat. A peacetime conscription, this development was an outgrowth of the Korean conflict when many reserve components were thrust into combat duty untrained and unprepared.

Clyde, a college friend, told me about this new requirement. He urged me to join the local National Guard unit in Stevens Point with him so we could enter the new program together. We would receive required basic combat training and advanced individual training in an assignment specialty within our unit. Enrolling in a local, hometown unit would be a six-year commitment but would exclude us from potentially being drafted. We could complete the required training over the course of one school semester and summer. Once we completed the training, we could return to our college studies. The chances of being called up seemed slim.

As I thought about this proposal, I realized that I was in a serious rut. While the summer at Camp Manito-Wish had put some spring into my step, the fall college semester felt laborious. I knew that I had a valuable military background. Enlisting with the National Guard would allow me to use these skills. After thinking about this option, I realized this might be another bugle call.

At the very legal age of eighteen, I once again joined the National Guard. Clyde and I enlisted in January 1956. We proceeded immediately to basic training at Fort Leonard Wood, Mo. This location is also known as "Fort Lost in the Woods" and "Little Korea." We quickly learned that both names had some validity.

## First Stop: Basic Training

Basic training is a challenge for everyone. It is not designed to be easy. As the entry point for soldiers, basic training is intended to put enlistees into situations where they are pushed to new limits. This time, succeeding in basic training was easier for me than when I was in the Junior ROTC. The platoon competition experience put me at the head of the class. I challenged myself to stay there. I worked harder than I ever had before, and it paid off. At basic training graduation, I was recognized as the outstanding trainee in the company. I felt I had earned this recognition and wanted to prove I was a model soldier.

I was a little older with a few more experiences under my belt. Yet, I was still very influenced by those around me. I began smoking during basic training because it seemed that everyone else

HEADQUARTERS
THIRD BATTALION
RESERVE FORCES TRAINING REGIMENT (PROV)
Fort Leonard Wood, Missouri

6 April 1956

SUBJECT: Letter of Commendation

TO: Private Raymond G. Boland, NG 27 865 582
Company "K", Third Battalion
Reserve Forces Training Regiment (Prov)
Fort Leonard Wood, Missouri

Upon your completion of 8 weeks basic training, I wish to take this opportunity to extend to you my personal congratulations for having been selected as the outstanding Reserve Forces Act trainee of Company "K", Third Battalion, Reserve Forces Training Regiment (Prov).

You are to be commended for your attentiveness, demonstration of leadership, your studied application in training, and your high character habits as evidenced by performing in an exemplary manner the duties of Trainee Squad Leader from 16 February 1956. On several occasions you assumed the duties of acting Trainee Platoon Sergeant in an emergency and performed efficiently. While performing guard you were selected as Colonel's orderly on 21 February and 23 March 1956. Your score on the M1 qualification was the highest in the company, 212, Expert Rifleman.

Again I wish to extend to you my congratulations for a job well done and the best wishes of the Staff and myself for your continued success on your new assignment.

Paul W. Long
PAUL W. LONG
Lt Col., CE
Commanding

did. Smoking and drinking were often synonymous with Army culture in the mid-1950s. Cigarettes were issued free to all soldiers in small packages as part of our standard-issued C-rations.

Has Fort Leonard Wood changed much since the days I was there during basic training? Yes. Recently, I attended a conference there. During a tour of the facility, I saw the current installation and training facilities. Very few of the old, World War II buildings I lived in remain. The post is now a world-class, fully modernized, total force training center. Technology is used to enhance training and is unlike anything we could have envisioned years earlier. I found this all very impressive and much improved from my basic training days.

One thing that has not changed is how soldiers are prepared and trained. Whilc touring, we encountered trainee formations using both the quick-time and double-time chanting cadences; exactly the same cadences I was trained to use 60 years earlier. I was not able to specifically measure the length and timing of the marchers, but they looked pretty darn good to me. I am confident Master Sgt. Mancini from my Junior ROTC training would be proud as well.

As I watched people train in their formations, the reality was not lost on me that these young Americans would serve their country much like the thousands of other people who have gone through basic training at Fort Leonard Wood. While they are going through the same building block processes that I and so many others have, the specific tools they use will be different. They have access to much more technology than we could ever imagine. Even basic provisions provided to soldiers, such as clothing, boots, and weapons, are radically different.

Yet, they will be trained on how to fight and win. Many have enlisted because they felt a bugler's call in their life, even if it came to them in a quite different manner than my call. These young people will protect and defend our country for the next generation. Their presence and abilities make me immensely proud.

## Next Stop: Fort Sill

After basic training, ten trainees from our regimen were sent to Fort Sill, Okla., for advanced individual artillery training. We were supposed to attend a radio repair course because this was the assignment for our home unit. Instead, we arrived to an unorganized fiasco.

At Fort Leonard Wood, a special reserve forces training regimen had been established just for the National Guard enlistees. Fort Leonard Wood brass was prepared for us when we arrived. When we arrived at Fort Sill, the people in charge had no idea why we and other National Guard trainees were there. The Artillery School was at a loss for a solution. Not sure what to do with these unexpected trainees, the leadership at Fort Sill chose to do the easiest and most expedient solution. They assigned us to line units at Fort Sill for on-the-job training (OJT), which would fulfill our required six months of training. They would put us through the ropes and get us out of their hair as quickly as possible. We would be assigned to the least desirable jobs possible.

A clumsy protocol was quickly thrown together. All the National Guard students were told we would take a written test. The forty people with the highest scores would stay at Fort Sill and the others would be eventually reassigned. Our entire gang of ten trainees from Fort Leonard Wood failed the test. Clyde, another new friend from Chicago named Mort, and I were assigned to the same unit: a 105 mm self-propelled artillery battalion.

When we arrived at the battalion, we were greeted with complete disgust. The first sergeant made it perfectly clear to the three of us that he had no time to babysit weekend warriors for two months. Our time in the battalion became nothing short of ugly, and it felt like a two-month replay of my two weeks at Camp Ripley—field duty, KP, guard duty, and serving as latrine orderlies comprised our training schedule. In reality, we were relegated to a special duty roster, and the lowest level you could be assigned to was permanent latrine orderly (PLO). I was threatened with this multiple times but never achieved it. One day, Mort actually told the first sergeant that he should know that Private Boland was an outstanding trainee in our basic training company. The first sergeant did not receive his curt remark well and almost landed me the PLO job for the duration of our assignment to this battalion.

Discrimination in the unit did not end with being called weekend warriors. As National Guard trainees, we received firsthand education about the chemistry between draftees and volunteers, or "lifers," as some guys called them. Active enlisted serial numbers began with the letters RA or US. USAR is the acronym for United States Army Reserve. As National Guard members, our serial numbers started with the letters NG. It was clear to my friends and me that NG was equated to simply mean "no good."

No love was lost between the respective categories within our battalion. Draftees with a similar enlisting code definitely tended to hang together. In general, I did not see particular performance differences. We had all gone through a similar basic training course and were there to serve our country. While everyone tried to do their job, the differences in how we were treated were dramatic. I experienced a culture different from what I expected; to me, a soldier was a soldier. Clearly, the higher-ups did not feel the six-month wonders should be accepted like other recruits. There was a significant disconnect between what Congress had enacted and how this played out in the U.S. military.

This whole experience was one that I often reflected upon later in my career. When I moved into military leadership positions, I was challenged to ensure that I treated every soldier like a soldier, no matter what pathway they took to arrive within my command. I did not want to be a person who unilaterally discriminated against or discouraged other enlistees. If I ever had the opportunity to correct this wrong in my military training, would I be able to do so with less disdain and clumsiness than I experienced? Later on, I learned how training seats are programmed throughout the Army. Based on this information, there was no way trainees from the new reserve training program could have been absorbed as quickly as we were signed up. There were not enough seats available for rapid implementation. Thus, National Guard trainees were simply viewed as excess soldiers.

Occasionally, my friends and I had time away from Fort Sill, and we would venture into downtown Lawton, Okla. Downtown was a conglomeration of beer joints where soldiers let off steam. Throughout my military career, I returned to Fort Sill three more times. By then, the downtown area had undergone a significant transformation. It became a place where soldiers could spend time together in a much healthier environment.

When our time at Fort Sill was completed, we left with no radio repair training. I felt my time at Fort Sill had been a complete waste of time. Despite the mistreatment, misuse, and disappointing reception into the battalion at Fort Sill, I still found myself interested in the Army as a whole. I tried to remember that the Army was bigger and better than my one awful experience. Maybe, just maybe, the Army could be for me.

## Assistant to the President

When the required six months of active-duty training were completed, it was time to put Fort Sill in the rearview mirror and return to Wisconsin. Clyde and I returned to Stevens Point, where I expected to resume college studies and continue training with the local National Guard unit. This was the original plan, but I encountered another bend in the road instead.

I found out that my family had reached a desperation point, and the motel cabin building project had stalled due to a lack of funds. My family was officially broke. My dad had managed to acquire some short-term work on a highway construction crew. This was due to end. My parents told me they could not financially help me attend school. Instead, they needed me to contribute money towards our family's survival. I knew that it was difficult for my parents to ask for help. My only option was to assist them. This was not a time to feel sorry for myself. School could wait, but my family's survival could not.

A new plan was implemented which included Dad and I relocating temporarily to Milwaukee, Wis. We would get jobs there and would come back to Friendship on weekends as often as we could. Marylu was now four, and my brother Al was sixteen. Dad and I got into my incredibly old and tired 1941 Ford two-door sedan. We headed for the big city of Milwaukee, which was about 150 miles away. The clutch in the car was so badly worn that we crawled up the uphill stretches.

Once in Milwaukee, we found a cheap room to rent. Then, we scoured the local newspaper's employment ads. Within two days of our arrival, we both had jobs and a place to stay. Dad's postal experience helped him land a decent-paying job with the Railway Express Company. My one-year college studies helped me interview for a job at Johnson Service Company, now known worldwide as Johnson Controls. The newspaper ad description was "Assistant to the President." The title sounded so important. I assumed it would offer many opportunities. As an outstanding trainee at Fort Leonard Wood, I felt qualified for the position.

The human resources person greeted me in a very friendly manner. It was a refreshing change from my job interview with the first sergeant at Fort Sill. After I filled out a detailed application, I waited for her to review it. Soon, she called me into her office. My background was suitable for the job. The starting salary was $55.00 per week. The compensation was significantly more than the $78.00 per month I received while serving as a private in the National Guard. I was overjoyed with this offer and confident it would lead to the executive position I yearned for. I assured myself this woman saw my true potential.

When I asked about the specific position, she said I would be working in their time study department. After she explained the duties involved, I knew I could manage them. Then I said, "Excuse me, ma'am, but I wanted to apply for the position in the newspaper listed as 'Assistant to the President.'"

She assured me that I had applied for the right position because, "Here at Johnson Service Company, EVERYONE is an assistant to the president." Yes, my 18 year old self was still very naïve.

I took the job. In a reality check moment, her words quickly registered with me. I felt relieved to have a job and the ability to contribute to my family's finances. I started working the following Monday, and the job was interesting. I enjoyed my coworkers, and they were genuinely nice to me. Each payday, I turned over my earnings, minus a few dollars, to the family survival fund. I had no regrets about doing this. In fact, I felt good to help in such an important way. I had little social life, but this was not new for me. With the combined earnings between my dad and me, our family stayed afloat. However, winter was coming, and heating the old farmhouse would be awfully expensive. We needed to adjust our plan soon.

My dad and I did not go home for Thanksgiving because we had to work Friday. It was a sad but necessary decision. We were invited by one of Dad's coworkers for Thanksgiving dinner and we appreciated it very much. Occasionally, Dad and I went to a local tavern for a beer. Hispanic guys wandered through the tavern selling tacos. It was our first experience with Mexican food. The vendors carried their offerings in baskets. They sold hot or mild tacos and tamales. Dad ordered us hot tacos, and we got scorched! They were fiery but very tasty. From then on, we stayed with the mild tacos. To this day, I love eating tacos and often think of Dad as the first bit of spice touches my tongue.

My dad and I went home for Christmas and had a family

conference. My parents wanted me to continue working, but my mother wanted me closer to home. She needed my help. I picked up a newspaper from the city of Wisconsin Rapids, which was 25 miles away. I found a posted position that included responsibilities similar to my position in Milwaukee, although it was not advertised as an assistant to the president. It paid the same weekly salary. I immediately applied and was accepted. Dad worked in Milwaukee for another six months.

My new position was for a company called Preway Inc. They manufactured kitchen appliances. My job was managing the inventory of hardware and other items needed to build the appliances. The working conditions were good, the employees pleasant and I was happy working there. The president was a very hands-on guy who made it a point to know all the employees. Ironically, I actually FELT like an assistant to the president of this company.

## A Yankee

When the summer of 1957 rolled around, things were looking a little better at home. The budget was still very tight, but there was food on the table. I felt comfortable with my job at Preway, and for the first time in a long time, I had a little time for recreation and enjoyment. My love for baseball came to mind, and I sought an opportunity to give it another try. During the first semester that I studied at Stevens Point, a collegiate baseball program had begun. Earlier, I had tried without success to make the basketball team. I had been playing intramural basketball, and now, the same coach was in charge of baseball.

I became friends with George Roman during my short stint attending college. George was also on the baseball team and urged me to try out for baseball. I did not feel my skills were good enough, but he was a very convincing guy. George knew the team was short of pitchers and said I should go for it.

George was an experienced outfielder, a good hitter, and a stalwart on the undefeated football team. In those days, he was the type of guy we called a "tough stud," and I was proud to be his friend. A very warm-hearted guy, George had taken the immature and lost me under his wing. He was so very kind to me. After a couple of weeks of practice, there was an inter-squad game. In the middle innings, the coach put me in, and it was disastrous. I did not get anyone out, and the other team scored about ten runs.

After the game, the coach pulled me aside to assess my performance. I expected the worst. Instead, the coach tried to make me feel good. He said, "Boland, the way I see it, you have three pitches." After his initial statement, I started to stand straighter. I fought back a grin and thought he assumed my pitches were a little wild in this outing. Instead, he explained. "The three pitches I saw today were slow, slower, and stop," he said. "I don't think we can use you."

George had the reputation of being a jokester, and he thought the coach's comments were hilarious. I now realized that George had set me up; he wanted to have some fun with me. He thought I was a cocky kid from Chicago and wanted to put me in my place, and he did.

A year later, I tried out for a baseball team again. This time, it would not be as a pitcher. In the 1950's, most Wisconsin small towns had a baseball league. Some towns still have baseball or softball leagues today. These games are fun and an important part of the local social culture because there are not a lot of entertainment options in these small towns. We lived in the community known as Adams-Friendship. Their team played in a league with other towns in central Wisconsin.

I did not know many people in our local community because I had been gone so much. I contacted the fellow in charge of the Adams-Friendship team. He asked if I could come to a practice and show my skills. I thought it was best to be honest about my abilities. I assured him that I was highly enthusiastic but limited in skill and experience. With few backup players, he thought I might be able to help the team. After working out, he said I showed more ability and potential than I had described. In order to get on the official roster,

I had to be cleared by the league officials. This way, teams could not slip unregistered "ringers" in at game time. A week or two later, I was eligible to play, but unfortunately, I missed the next couple of games due to work and National Guard meetings.

A couple of weeks later, I came home from work one day. My mother was extremely excited and could not wait to tell me something particularly important. She started, "Raymond, I had no idea you were such a good baseball pitcher." I confirmed with her that I was not a good pitcher but, in fact, a lousy pitcher. She proceeded to tell me that a man had been at the house that day looking for me. He was a scout for the New York Yankees. He confirmed that I was a really good pitcher and had recently pitched a one-hitter and a no-hitter in successive games. He wanted to talk to me right away.

Initially, I had no idea how this strange case of mistaken identity happened. But then, the proverbial light bulb came on. After one phone call, I confirmed that the team had used my name for someone else they had recruited to pitch. He was an outstanding player who had just graduated from Wisconsin Rapids High School. He was an unknown face to the teams in our league. I was also an unknown face, so, they simply put the registered name of Ray Boland in the official lineup. As they say, the rest is history. When the whole scheme unraveled, the Yankees found the real guy and signed him to a minor league contract with a substantial bonus. Four years later, he was a relief pitcher for the 1961 Yankees World Series championship team. Soon after that, he developed arm trouble, and his career was cut short.

Naturally, the local team had to forfeit the games they had won and maybe even the entire season. Once again, Raymond faded back into obscurity. Today, this story is one of the more humorous anecdotes in my portfolio. At the time, it did not feel very funny at all, and my mother's new hopes for me to become famous rapidly fell apart. I lost all interest in playing baseball as the real Ray Boland. I had no desire to show up at a field, confident I would be the butt of everyone's jokes.

There was one upside to this story. I called my friend George in Stevens Point to share with him the good news. I wanted him to be the first to know that all the faith, trust, and confidence he had placed in me had paid off. Had it not been for his encouragement to become a pitcher, my true potential might have never been recognized. Now, the New York Yankees wanted me to pitch for them. He should be proud of inspiring me and relay my success as a baseball pitcher to his friends.

Many years later, the Wisconsin Rapids newspaper ran a feature article about the "imposter" baseball pitcher and how the New York Yankees originally discovered him. The real pitcher accurately told the story of how he was recruited to pitch for my local team. He shared how the team had used the "fictitious" name of Ray Boland to get him into the game before he was approved for the roster. When I read this article, I felt like I had hit an all-time low. I had gone from obscurity to now pure fiction.

The final footnote of my baseball saga happened when I attended the funeral of my good friend George. During his celebration of life gathering, I shared the story of my dismal baseball team experience. Everyone agreed this was a classic George caper. He always had a way of finding a place to occupy your heart. George has stayed with me all these years and is one of those people who is not easily forgotten.

I knew the call to play baseball had officially ended. Would there be another call into active military service? At this point, I did not anticipate this. But fate has a funny way of entering our lives, as I would soon learn.

# Chapter 5

# Back to College

As the summer of 1957 drew to a close, things continued to improve at home. Dad was hired as a deputy sheriff for Adams County. He stopped working in Milwaukee and was home. The first two duplex cabins were completed. The motel was officially open for business under the name the Prestonaire Motel. Everyone my parents ever knew in Chicago came to visit, which helped create a customer base. My job at Preway was going well. I had some new friends in Wisconsin Rapids, where I found things to do socially.

My thoughts began to drift toward the idea of returning to school and completing a college education. I still could not afford to go back to Stevens Point. During the previous year, my earnings had helped support our family. I had not been able to save any money for school.

One day, I saw an item in the Wisconsin Rapids newspaper that opened another unexpected door of opportunity for me. It was an announcement for student enrollment and admission procedures for Wood County Teachers College.

I was unaware this school existed. It was a two-year school

in Wisconsin Rapids devoted entirely to teacher training. Their curriculum qualified graduates to receive a two-year conditional teaching certificate from the Wisconsin Department of Public Instruction. Graduates could teach in any Wisconsin public elementary school. The two-year certificate was a long-standing teaching option aimed primarily at providing teachers for one-room rural schools that taught all eight grades. These colleges had previously been called normal schools and the state opened several of these colleges in the early part of the 20th century. In 1957, a number of normal schools were still open in Wisconsin. Just a few years later, they were all closed.

At the time, each local public school district set its own hiring policies. Many larger school districts also hired two-year certificate holders. State policy required that in order to remain certified, continuing education toward a four-year teaching degree was required. Additional credit hours were necessary within prescribed time periods.

If I enrolled at Wood County Teachers College, I could commute daily to Wisconsin Rapids from home and work part-time to make ends meet. The tuition was much less than the state college and small loans were available from the county. If the school accepted some of my credits from Stevens Point, I could finish in less than two years and start teaching. This was not what I hoped for, but an option that headed me toward the direction of a four-year degree. I made this choice as a “Why not? What have I got to lose?” decision. I enrolled and one semester of equivalent credits transferred from the state college. I could complete the program in three semesters and become a qualified teacher.

When classes began in September, I was not sure what to expect. I was far more ready to be a serious student than two years earlier. Yet, I was curious about how I could become a teacher in such a short time. The school was small. I was surprised to discover that the total combined enrollment of first- and second-year students was only sixty. There were five faculty, including the president, Mr. Warren Lensmire. His wife was the single administrative staff person.

The students and faculty were a very close-knit group. They worked together intensively to train students to become teachers. This was the reason students attended this school. This is what we did. The school had its own full-time classroom of students in grades one through eight. They provided us with a place to practice what we learned. Prior to graduation, we practiced teaching at individual schools in the area under the supervision of the school’s regular teachers. This was hands-on training at its best, allowing students to quickly develop confidence that we could teach.

The curriculum was very focused on how to teach and how students learn. Reading and reading readiness were the foundation of everything we did. We were fully ingrained with the concept that children must be able to read in order to learn. It is hard for me to believe that some of today's school-age kids cannot read. In 1957, most school budgets, especially rural schools, were very austere. There were no technological teaching aids. Audio-visual aids included a film projector or public education radio. The chalk board was our primary teaching tool. Yet, students learned. We were required to administer standardized tests at the start and end of each school year to measure progress. In spite of comparatively major increases in funding over the years, I feel today's Wisconsin K-12 education has made questionable progress in the past 50 years. In some respects, educational learning may have regressed.

Wood County College's only extra-curricular activity was intercollegiate basketball. It felt fortuitous that this was the only option. Each of the county teacher's colleges had men's and woman's teams. I joined the men's team, and we played a regular season schedule and concluded with a tournament. Finally, I became a starter and was the leading scorer on the varsity team. Should I mention there were only seven players?

For the very small numbers of participants, the competition was decent, and the games hard fought. In my final game before my mid-year graduation, I scored a lifetime high of 39 points in a game. I hit deep jump shots throughout the game against a zone defense. At one point, I downed ten shots in a row. This was before the advent of the three-point line. Had a three-point line been part of the game, total points would have been more.

The opposing team stayed in the zone and conceded my shot until they switched to a double-team, man-to-man defense for the game's last couple of minutes. I found open teammates with passes, which kept us in the game. I was fouled with only six seconds left in the game. We were ahead by one point. I was an 80% free throw shooter but choked on the front end of a one-and-one. The other team rebounded the missed shot and made a sideline outlet pass to a guy who took one dribble and threw it up from half-court. It was all net at the buzzer. We lost by one point, 70-69. My college basketball days ended, embracing the well-known saying, "It ain't over 'til it's over."

## Love and Marriage

When I started school at Wood County College, I met a student named Jackie Parker. Instantly, I knew she was the someone I had been waiting for. I was convinced our relationship was meant to be. Unfortunately, I did not make a very good initial impression on her because she thought I acted like a wise guy who only wanted to impress her. I concluded that she was very astute and a quick study. She was all of this and so much more. Eventually, I spent the next 46 years with her.

Before she was ready to commit to a life-long relationship, I had to first dig myself out of the hole from my botched first encounter. I came up with an unusual plan and knew it would be tough to pull it off. But I was determined to make it happen.

I made my move on the day of the second regular basketball game of the season. The game was away. The team, faculty, and student body rode on a bus together to the game. One of my teammates, Dick, had a scheduled date with Jackie after the game. I had a planned date with another girl. The scheme included convincing my teammate that my date was crazy about him and wanted to date him in the worst way. I told him that I was willing to switch dates and take Jackie off his hands, which would allow him to see the other girl. He bought it.

Next, I had to convince my date that Dick was crazy about her. I told her that Dick had asked me to step aside for the evening and give them a chance. She bought it. So far, so good. The real challenge came when I had to sell all of this to Jackie. It was not easy. At first, I thought my well-thought-out plan was going down in flames. Her opinion of me seemed to quickly shift from smart ass to dumb ass. (These were not her words but my interpretation of her reaction.) It took my most daring act to close the deal.

I told her the truth.

I held my breath as I waited for her response. She started to laugh and said she would give me style points for the most outrageous line of BS she ever heard. Jackie agreed to the plan. From that night forward, we never parted. Dick and I remained lifelong friends. He completed a career as an elementary school teacher. Dick and the other girl hit it off. They dated for quite a while but did not marry each other.

Dick was also instrumental in connecting me to the Society for the Preservation and Encouragement of Barbershop Quartet Singing in America. We both enjoyed the close harmony and fellowship among barber shoppers. We had some good times with them. He was also a member of the National Guard, which becomes another connecting point I share later.

Jackie and I were from very different backgrounds. She grew up in a very small town while I was from the big city. Yet, we shared so many similar thoughts and ideas. We did not agree on everything, but we saw the world in much the same way. We both felt quite a bit of uncertainty about our futures.

Jackie's parents were very supportive and pleasant. Yet, she was ready for something beyond the smalltown boundaries where she grew-up. She was ready to explore the world. Likewise, I was searching for a new direction. Neither of us knew exactly what we were looking for. We both ended up at Wood County College because we did not know what else to do. Our mutual desire to find something new in life drew us together. Quickly, we became inseparable. With each passing day, my admiration, love, and respect for her grew stronger.

In a few short months, we were married. We knew it was not the most practical thing to do. Our relationship was based on the belief that together, we could accomplish more than either of us could on our own. Jackie was very mature, talented, and caring. I was extremely fortunate to marry her. It was the best thing that could have ever happened to me.

## Wisconsin Military Academy

While attending Wood County in the fall of 1957, I continued my required weekly National Guard unit meetings in Stevens Point. One night, the first sergeant instructed me to report to the battery commander. He could not tell me why this was requested. By now, I was a specialist four (Spec 4). I thought I was doing a good job in the unit. After returning from the six-month basic training course where I was recognized as the outstanding graduate of my basic training company, the chain of command deemed me a good soldier. Every soldier knows seeing the commanding officer (CO) is seldom a good thing. I assumed that I had done something wrong and was nervous about the meeting.

When I came before the CO, I used the proper reporting procedure and rendered my best hand salute. The CO asked me to stand at ease. He proceeded to commend me for the fine record I had achieved since joining the unit. Then, he told me something I completely did not expect. He wanted to recommend me to become a commissioned officer. He explained that the National Guard was implementing a new program nationwide to train officers and would be called the State Officer Candidate School (OCS). In Wisconsin, it would be known as the Wisconsin Military Academy (WMA) and would begin operation at Camp McCoy in June 1958. Nebraska was the only other state with a similar program in place. Wisconsin would be the second state to establish this type of program.

Previous to the beginning of the WMA in Wisconsin, National Guard officer training was completed mostly through correspondence courses rather than resident schooling. Later, I came to understand this new program had the very same objective as when I joined the Wisconsin National Guard unit. The intention was to raise the training standards for all National Guard enlisted soldiers and officers.

It was a one-year training program, beginning with a two-week annual training phase at Camp McCoy. This would be followed up with one weekend a month at the state National Guard headquarters in Madison, followed by another two-week phase at McCoy. Class #1 of the WMA was scheduled to graduate in June 1959. Graduates would be commissioned as second lieutenants.

I was surprised and a little stunned to hear the CO's

recommendation. My immediate reaction was that I did not want to become an officer. I only intended to complete my six-year military obligation and be done. He asked me to think about the possibility before deciding. Unfortunately, there was not much time before applications needed to be submitted. He suggested that we talk again the following week and discuss any questions I had.

As I drove the 50 miles home to Friendship, I could not stop thinking about the WMA opportunity. The thought of becoming an officer was not entirely new to me. Master Sgt. Mancini's words from my early Junior ROTC program about being a leader rang in my ears. I contemplated my days as a private and how other privates and I were treated by those with higher ranks. If I became an officer, I would have the chance to be the kind of leader for whom I had only wished. The reality of what would be required seemed more than I wanted to add to my current plate of responsibility; another dilemma.

Several times earlier in my life, I had made a big decision. This time, the stakes felt bigger. This seemed like a much more significant fork in my life.

So much of life's journey is about making decisions. Often, they are very hard to make. There are moments in our lives when we must reach deep inside ourselves and depend upon our inner faith and judgment to discover the best solution. I decided to confide in my father about this option. We had a helpful discussion. Dad urged me to seize the opportunity and make the most of it. He shared that if my unit was ever called to active-duty, it would be helpful for me to be an officer rather than my current rank as an E-4. His reasoning did not factor much into my decision, and the likelihood of being called up seemed extremely small. With World War II and the Korean conflict behind the U.S., I did not anticipate there would be another war any time soon. In four short years, I expected to be draft-exempt.

Yet, Dad understood the practicalities of being an enlisted soldier versus an officer. During World War II, Dad served as an enlistee. He rose to the E-4 and E-5 ranks, and he understood the challenges of being an enlisted soldier during wartime versus a technical specialist.

Dad's closing argument influenced me the most. He said if I went through the training, he was confident I would never regret this decision. Achieving officer status was an achievement that could never be taken away from me. No matter what else happened in my life, this title will always be mine to keep. Little did we know at the time that my father's words would become a prophesy beyond his imagination.

It was a lot for me to process, and I felt overwhelmed with all the decisions I needed to make in a short period of time. A pending marriage to Jackie, becoming a teacher, moving several times, and now, another big decision. I yearned for harmony in my life and confidence in the work I pursued. I wanted a sense of purpose in my life so that I could succeed and support a family. My strong sense of faith gave me the necessary confidence to work through these tough decisions.

I returned to training the following week and informed my CO of my decision to apply for OCS admission. The CO coached me through all the paperwork and said it would take a while for it to be processed. A selection board screened all the applications. At some point, I would appear in person before the selection board for their final decision. Only then would I be informed if I made the cut and was accepted into the program.

## Wedding Bells

Jackie and I decided to get married on a Friday evening after the completion of a regular school day. We delayed our honeymoon because classes resumed on Monday. I was only 20, and Jackie was 18. Interestingly, she was of legal Wisconsin marrying age, but I was not. I needed parental consent to purchase a marriage license. My mother was reluctant to do so. She told me that if I took this step, I would never amount to anything. I should no longer expect any help from her.

Her response was a classic, "You made your bed; you lay in

it" declaration. This was a very common societal attitude in those days. She had such high hopes for me and felt that I was headed in a direction where I would not achieve this. After some pleading, my mother agreed to sign for me. I was so upset that I pledged to myself that I would prove her wrong and my marriage *would* work out. I had to buckle down and get serious because someone else was going to be dependent upon me. Years later, I often came to realize that this marriage forced me to mature and grow up. With Jackie's support, I was able to accomplish so much.

Jackie and I arranged a small wedding. My brother Al was the best man. A college friend of Jackie's was the maid of honor. Jackie's father hosted a dinner reception at a local supper club after the ceremony, followed by a casual get-together at my parent's home to complete the evening. All the necessary plans were in place.

Two days before our wedding, I received a call from the National Guard headquarters in Stevens Point. The OCS selection board planned to convene on Saturday in Stevens Point at 8 in the morning. This would be my in-person interview. I pleaded with the person on the other end of the phone, "But I'm getting married Friday night!"

The person said, "Congratulations! Be at the headquarters Saturday morning." If I did not appear before the selection board, my OCS application would be withdrawn. There were no second chances.

Jackie's experience with the military only included her brother serving during World War II. She had no idea what OCS was. Jackie and my dad quickly hit it off and became fast friends. Both Libras, their goal in life was to make everyone happy. I told them that I was upset with the timing of the interview and had decided to be done with my OCS application. They both assured me that I could make the selection board appointment on Saturday morning and do just fine. Jackie deferred to my dad's assessment that this opportunity was very important. She felt it was something I needed to do. Jackie and I decided that we would retire early Friday night so I could rise and shine early Saturday morning for the appointment.

Well, our plans went awry, as the quiet gathering at my parent's house turned into a raucous party. Yes, the group was small, but boy, everyone had a good time! Those in attendance included my dad's boss, the county sheriff and his wife.

It was late when my brother asked to borrow my car to take his girlfriend home. Shortly after they left, he called, as the transmission had just gone out on my 1950 Ford. My dad did not own a car. Like the sheriff, they just drove their police cars. For the second time in two days, I felt like throwing in the towel. I was becoming convinced that the OCS would not be part of my future. Just when everything seemed to have fallen apart, the sheriff's wife offered me her car. As the dispatcher and jailer for the small sheriff's department, she was scheduled to be on duty and would not need her car.

This is another time when my life would have taken a very different direction had the sheriff's wife not offered me her car. I thanked her and promised that I would be very careful with her car. Sometime after midnight, we all finally went to bed. Jackie and I spent the night at the Prestonaire Motel.

At 0800 hours the next morning, I was at the Stevens Point National Guard headquarters, ready to go before the selection board. I was tired, groggy, and wishing I was still at the motel. Several other applicants were also there. One at a time, we were called to report to the board. This exercise was much more comprehensive and challenging than I expected. Quickly, I realized the board took their responsibility seriously. No softball questions were tossed my way. They gave close attention to our appearance and military bearing, which relates to a soldier's image and demeaner, knowledge, and manner. I was even required to perform a series of facing movements. It was a true workout in every way possible.

After the board had completed their interview, the board president directed some less-than-positive feedback my way. He said, "Specialist Boland, I want to commend you for the outstanding record you have achieved during your service. Your record shows exactly the kind of potential officer we are looking for. You meet the high standards required to become an officer."

The accolades stopped there, as it was puzzling to him that the person I presented to them physically was not the same person they observed on paper. He asked me if I was aware that I did not have a belt on my trousers. Our issued uniform included the World War II era Eisenhower jacket that came only to the waist. Was I aware that I was wearing two different socks? Only one of them was military-issued, and my answers to their questions seemed sluggish. In summary, I was not what they expected to see, and he asked if I had any explanation for these shortcomings.

I responded, "No excuse, sir." I followed this command with the simple truth that I had gotten married the previous night and admitted that I was not as prepared as I should have been. He responded by commending me for leaving my new bride to report early this morning. This demonstrated to the board the kind of strong commitment the National Guard felt was necessary to become an officer.

As the officer excused me, he offered their congratulations on my wedding and extended best wishes to me and my bride. He also directed me to report directly back to her.

As I drove those now very familiar fifty miles back to the motel, I thought to myself, "Good try, Ray, but you blew it."

Imagine my surprise when, two weeks later I was notified I had been accepted into the OCS program. My first conclusion was that getting accepted into the OCS program was meant to be. Now, I began to believe that I was not only called to be a soldier but also a leader of soldiers. This was a new responsibility for me and would be unlike any other training I received.

## Officially, Mr. and Mrs. Boland

Yes, I did take my new bride on a honeymoon. We went to Niagara Falls. How did two college students afford this? Easy. We participated in a school trip to Washington, D.C., with our fellow classmates. It was a bus trip held at the end of the school year. We had fundraised throughout the school year to pay for the trip.

There were no bells or whistles or flashy extras on this trip. It was about as low budget as you can get. We stayed several days in Washington, D.C., at a place called the Potomac Park Motor Court. There was another married student couple on the trip, and we stayed with this other couple in the same room. One of our visits to the nation's capital included meeting with our Congressman, Melvin Laird. From Wood County, Melvin served as Secretary of Defense during the Vietnam War.

After college was out for summer break, we lived in a small

upstairs apartment in Friendship, which was the town closest to my parent's motel. I worked at a gas station and began officer training. Phase 1 of OCS went well. I was very impressed with the program's quality, as well as the commandant and tactical officers. These officers were very good trainers and extremely professional. While at Fort Sill, I heard about the rigors of OCS. I was prepared for the training to be tough, and it was. Several candidates did not last beyond the first two weeks.

For me, the most difficult part of the course was the Madison weekend training sessions. I continued to work while attending school. We lived in another upstairs apartment in Wisconsin Rapids, just minutes from school. Our rent was $25.00 a month, and my weekend hours were precious and valuable. More than once, I wanted to call it quits. It would have been much easier to just work than also attend officer school. I remember one particular day when Jackie literally pushed me out the door.

In November 1958, we were blessed with the arrival of our first son, Daniel. It was the proudest moment I had ever experienced. Daniel was good-sized, a very strong and healthy baby. Right away, everyone called him Danny. That is, except Grandpa Parker, who called him "Boone." Immediately, I saw in him a future high school basketball player. I anticipated living out my basketball dreams through my son, and in time, Daniel earned a spot as a varsity basketball player and was a very skilled football player. By the time he was in high school, our family lived in Texas and Oklahoma, where football is king.

My graduation was scheduled for the middle of the school year. Historically, this is a difficult time to land a teaching job. I resigned myself to the fact that I would work full-time at the local supermarket and pick up part-time work to support our young family. Once again, someone advocated for me, and a wonderful opportunity landed in my lap.

During my tenure at the college, I grew to respect and appreciate Mr. Lensmire, president of the Wood County Teachers College, very much. Mr. Lensmire was a very dedicated educator for our small college. His enthusiasm was contagious. Mr. Lensmire made you feel like you could be a great teacher. In his mind, the most noble thing a person could do was be a teacher.

Mr. Lensmire was a personal cheerleader for Jackie and me. He expressed great confidence in our future together and showed trust in my abilities when he approached me just a couple of weeks before graduation. An unexpected teaching position had opened up due to a resignation. He recommended me for the job. The position included being principal of a three-room school in tiny Babcock, Wis., where I would teach the sixth through eighth grade students.

Mr. Lensmire was honest with me, as there had been some behavioral problems at the school. The previous teacher had resigned because of these unruly students. Yet, Mr. Lensmire was confident I could handle the situation and turn the classroom around. The minor detail of my needing a principal's license in addition to my regular teaching license needed to be worked out. But he would get it all approved. With Mr. Lensmire's blessing and encouragement, I formally applied for the job, was interviewed and hired. The starting salary was $335/month. While the responsibilities would be much more demanding than the supermarket, this was big money for me. I would be able to support my family.

## Mr. Boland, the Teacher

My first day as a schoolteacher/principal was a typical cold and wintry January day. It reminded me much of the day when I first arrived in Wisconsin just a short three years earlier. I paused to think about all that had happened since arriving in the central sands of Wisconsin. The list was long: Central State College in Stevens Point, enlisting in the Wisconsin National Guard, attending training at Fort Leonard Wood and Fort Sill, working with my dad in Milwaukee, working in Wisconsin Rapids, attending Wood County College, and being accepted into the OCS officer training program. My biggest successes? Jackie and Daniel. Now it was time to prove that I could succeed in being in charge of something. I was nervous but could not let the students see this, which would be a disaster.

Before the students arrived, I wrote my name on the chalkboard and greeted each of them as they entered the classroom. I was aware these students lived in a very low-income community with a very basic and down-to-earth lifestyle. It reminded me somewhat of my old Chicago neighborhood. I made no reference to the rumors of their conduct or the departed teacher. From the very beginning, I treated the students as if they were the best group of kids. I let them know that I was proud to be their teacher. I greeted them every day like it was going to be the best day we ever had. I wanted school to be a highlight of their day and not something they had to do.

Most importantly, I committed the effort to allow every student to feel a sense of achievement. During Army basic training, I learned that achievement is a process of building blocks. Every student needed to begin at the level they were at. Every small baby step achievement could lead to a larger accomplishment, and so on. The kids responded well to this training, and we really had a great time working together. I discovered that these were really good kids.

Shortly after I began, I asked the boys if they liked basketball. Several students raised their hands, and I proposed that we start a school basketball team with the goal of playing against the big school up the road called Pittsville. The boys were ecstatic about this possibility but quickly pointed out there was no gym for practice. I told them not to worry because there was a basketball goal on the playground behind the school, so we could practice there.

The students pointed out that the playground was buried in snow. By a show of hands, I asked how many of them had snow shovels at home. They all raised their hands. I reached into the Master Sgt. Mancini playbook and told the students if they came to school early every day with their snow shovels, we would clear an area to practice, and we would practice winning.

I purposely did not divulge that I had also cleaned a practice area behind our house in Chicago because my mom wanted me to commit to something. While they were not convinced shoveling was a very good idea, they agreed to give it a try. I made it perfectly clear that this plan could not be carried out unless they also worked hard in the classroom. With a deep desire to play Pittsville, they agreed to my conditions.

A frozen playground was not an ideal place to develop a basketball team. Nonetheless, I made the best use of the little practice time we had. From my own basketball experience, I learned that ball control, effective passing and ball handling were the most important skills they needed to learn. Dribbling a ball on our outside court became another teachable moment, and we worked hard to learn why and how each pass was made. Those who have seen the movie *Hoosiers* might think I took a page out of their playbook. Our training sessions were long before the film was made.

Finally, it was time to play our big game. What a game it was! The Pittsville team easily outmanned us, and our little country school team stayed in the game as it wound down to the last shot. Their best player had the ball in his hands and delivered the winning shot, so we lost a heartbreaker by two points. Our kids handled it pretty well, and they knew they had played well enough to have won. A substantial crowd of students and fans supported the team and showed up at the game. They were impressed with what our kids had accomplished. The buttons on my shirt almost burst from pride.

About two months later, the Pittsville superintendent of schools contacted me. He wanted to discuss a possible teaching position. At our in-person meeting, he asked if I would be interested in being their seventh-grade teacher. Along with this, he asked me to establish a grade school basketball program for fifth-grade boys and up. The job also included coaching the high school JV team, and a salary increase I could not pass up. Over the summer, he offered me a job painting rural school buildings. I was aware the incoming freshman class would include kids from both the Babcock and Pittsville teams that participated in our big game. Quickly, I accepted his offer.

## Lieutenant Boland

In June 1959, the Wisconsin Military Academy (WMA)

reconvened at Camp McCoy for the final phase of Class #1 as well as the initial phase of Class #2. As an upper-class student, my classmates and I were expected to demonstrate the highest standards to the incoming "plebe" class. After another challenging two weeks, on June 19, 1959, I was commissioned a Second Lieutenant, Field Artillery. Jackie, Dad, and my brother Al attended the ceremony. It felt like Dad was even prouder of my accomplishment than I was. Maybe his own dream had come true, and he was living vicariously through me. My unit assignment was to the "B" Battery, which was part of the 121st Field Artillery Battalion based out of Marshfield, Wis. I was commissioned as the Battery Assistant Executive Officer. This battalion was equipped with 155 mm towed howitzers, a step up from my previous experience on the 105 mm.

The state of Wisconsin bestowed our commission, which was followed by a required federal recognition from Washington, D.C. Another program requirement was to complete a resident Active Army Officer Branch Basic Course within three years of our commissioning. I needed to attend an eight-week course at the Field Artillery School at Fort Sill. I returned to Fort Sill wearing the real gold bar that Master Sgt. Mancini lectured me about it in high school. I was beginning to understand what leadership responsibility truly meant, but I had a long way to go to become a qualified field artillery officer who could perform those duties in combat if called upon. The joy of achieving the goal of becoming an officer felt very fleeting. Little did I know that I would carry the reality of those responsibilities for a long time.

Summer was over, and it was time to become Mr. Boland, the teacher/principal, once again. The 1959-60 school year was a wonderful experience, which I enjoyed. We were back in Pittsville, Jackie's hometown. We found a house to rent just two doors away from Daniel's, aka "Boone's," grandparents. The same school district hired two of our Wood County College classmates, and we had our own little alumni group in the same school district. Another Wood County College classmate replaced me at Babcock.

My seventh-grade students were ready to learn; likewise, the basketball season was a lot of fun. We lived in a small town in rural America, where the connections often run deep. Very deep. Later, in the early 1970's, my father-in-law moved into a care facility. He left the home he had built himself, the same home where Jackie grew up. One of my seventh-grade students and her husband bought the home. It is now some 50 years later, and she still lives in this home. When I visited the home, the house was much like I remembered it. My student, now retired, was as nice as I remembered her.

As the school year drew to a close, I was offered a new job opportunity with more responsibility and better pay at a different school district. I was offered the principal and eighth-grade teacher position at the Roche Cri Elementary School in Arkdale, in Adams

County. This was a consolidated six-room rural school with approximately 130 students. Jackie and I were not enthused about moving again, but we weighed all the necessary factors. I had started the required continuing education requirements toward receiving a bachelor's degree by attending Saturday classes in Stevens Point. I anticipated the day when a four-year degree would open the door towards a full range of teaching opportunities. I felt the strength and experience of my resume could influence potential future teaching positions. The Arkdale position would give me the opportunity to demonstrate more leadership and responsibility.

Another factor that crept into the decision was that my parents were moving back to Chicago. After struggling for five years to start their motel and survive financially, they accepted temporary defeat. Their new plan was to return to full-time employment in Chicago, where they would work and save. They planned to return to Friendship and resume their goal with the Prestonaire Motel. Of all the decisions I had watched them make in the previous fifteen years, this choice felt the most practical. My dad would be reinstated with the Chicago Post Office and planned to complete enough years of service to be eligible for a pension.

For the next 12 years, my parents lived in Chicago. They made the 250-mile trip to Friendship every weekend they could. Concerned about leaving their property vacant, they offered us the chance to live in it rent-free if I took the position at Arkdale. The

Prestonaire was only eight miles away from the school, and we made the move.

While we lived in Pittsville for only one year, the little burg of a town always held a special place in my heart. The memories were fond: time spent with Jackie and her family, other families and friends we became close with, visiting Riverside Park with my students, the annual Pow-Wow, basketball games, baseball games, and visiting the marker in town that indicated Pittsville as the geographic center of the state of Wisconsin.

Several years after I left Pittsville, my fifth-grade basketball players had a remarkable season as high school seniors, and they advanced to the sectional finals. In those days, the little schools played against the big ones during tournament season. Their coach shared with me about their valiant effort in a close loss. At the time, I did not know him, but in more recent years I have had the pleasure of playing basketball with him in the national senior Olympic games. Like me, he holds Pittsville in high regard.

Our move in the summer of 1960 was overshadowed by the arrival of our second son, David. Born in August, he was another husky little guy and very good natured. He looked like a football player from the beginning. Eventually, he followed his brother Dan's footsteps. David played tight end while we lived in Oklahoma.

That summer, I also attended the required annual training at Fort McCoy. I was now an officer and anxious to prove I was up for the task. On the first day of training, we moved into the field for tactical training. I was responsible for establishing the fire direction section at the initial location we were given. The battery commander let me do this on my own. Before departing, he gave me special instructions. An active Army major would be evaluating our unit, most likely visiting each of our positions on the first day of training. He would check to see how well we adapted to our field conditions. The battery commander conveyed how important it was to make a favorable impression. This particular major was known as a hard-nosed individual. My commanding officer (CO) handed me a copy of the prescribed Army Training Test (ATT) task list for our unit. This was the most important part of our evaluation. My instructions were to properly report to the major if he showed up, show him the task list and tell him, "Sir, today we are practicing for our ATT."

About 30 minutes later, the major appeared, seemingly out of nowhere. He came stomping through the woods. As he walked up to my section, I was kneeling on the ground, sorting through the poles and pieces for our tent. Exactly as directed, I called attention, saluted, and said, "Sir, today we are out here in the field practicing our ATT." I dutifully held up the list. He looked at me, looked at my soldiers, looked at the stuff on the ground, looked back at me and said, "Bullshit." He turned around and stomped back into the woods.

About an hour later, my CO came to our site, and we were still wrestling with the tent but making progress. Immediately, the CO quickly asked if the major had been there. I replied, "Yes, Sir."

The CO asked if I had followed instructions. I confirmed, "Yes, Sir."

Then, the CO asked if the major said anything. I said, "Yes, Sir, he said, 'bullshit.'"

My CO could not believe that I could screw up something so simple, which led him to also tell me that he did not think I had the potential to be an officer.

I made the decision this was the last time I would ever try to BS a senior officer.

In September 1960, I began a new school year in Arkdale, Wis., as the eighth-grade teacher and school principal. We had an excellent team of teachers and a modern school building, and it was a very pleasant setting for teaching and learning. Located in a rural farming community, most students rode the bus to school. I was a substitute bus driver and occasionally filled in.

Another basketball team was created without a gym, but we did have one indoor basket in the school lunchroom. During this school year, I realized I enjoyed coaching basketball more than classroom teaching. I wondered if this might be my true calling. I also found myself starting to lose patience with students who misbehaved. On one occasion, I completely lost my cool and became overly physical with an eighth-grade student who was bullying a fourth-grade student in the school hallway. When his father showed up the next day at school, I anticipated the worst. Much to my surprise, he thanked me for disciplining his son. He also asked if his son acted up again to call him so he could take additional action.

After the father left, I felt relieved. I was also keenly aware that many parents would not have reacted this way. I could have been in big trouble because I lost my cool. In today's environment, I probably would have been let go. I never wanted to be in this situation again, and it was time to seriously consider a different line of work.

When the school year ended, I was offered a new contract for the upcoming year, but I turned it down. My required military schooling at Fort Sill was scheduled to last through most of September. I would not be back in Wisconsin when the fall semester began. I had a heart-to-heart talk with the district superintendent. We agreed it would be best to find someone else for the position. He hired the same college classmate who replaced me at Babcock. He stayed in this position until he retired 30 years later.

As the summer of 1961 began, I was officially unemployed.

Even more challenging, I had no idea what I would do after completing Army school. The practical side of me struggled with supporting a wife and growing a family. I had obligations. Despite my questionable decision-making process, something was telling me it would all work out. I was confident I would figure it out and something new would come my way. Soon, this became a self-fulfilling prophecy.

# Chapter 6

# The Berlin Crisis

After the 1961 Arkdale school year ended, I attended the Field Artillery Officers Basic Course at Fort Sill, Okla. Jackie and the boys went with me. We rented a furnished apartment in an off-post complex. Almost all the tenants were military families. Summers in Lawton, Okla., are not exactly balmy, and the apartments were not air-conditioned. Instead, there were roof-mounted water coolers that provided some relief. For the most part, it was just plain miserable. David was just one year old. His heat rash was impossible to get rid of. Thousands of military students have lived in these kinds of conditions. It is simply how it was.

Almost all the students in our eight-week course were newly commissioned second lieutenants who had just graduated from college. Most had been commissioned through ROTC. A few were West Point graduates. I was the lone National Guard officer in our section. The standards were set high. I struggled to keep up with the challenging coursework. These officers had a higher quality education and were a more capable group than I was used to. The competition environment was daunting.

Fortunately, we had an outstanding Marine Corps captain as our gunnery instructor. He presented complex material in an understandable way. At the time, fire direction data was calculated manually because there were no computers. This task is necessary to provide aiming instructions to the gun crews as they are unable to see the target area, which is five to seven miles away. Forward observers (FO) radio back where the artillery landed. Anyone who trained at Fort Sill is familiar with the "blockhouse on signal mountain" reference point.

One morning, our gunnery instructor entered the classroom and announced, "Lt. Boland, congratulations! You are now a member of the United States Army." A potential crisis was evolving in Berlin, Germany. In response, President Kennedy had just activated the Wisconsin 32nd Infantry Division and the Texas 49th Armored Division, along with numerous other reserve component (RC) units. A stunning development! There was great uncertainty about what was going to happen. Only two things were clear: firstly, I would no longer be job hunting. Secondly, I would be going to Fort Lewis, Wash. Beyond this was anyone's guess.

My biggest concern in being mobilized was not for myself but for my family. This is a common reaction among RC soldiers. In contrast, active component (AC) soldiers are more aware that they may be deployed on short notice. Their families understand this as part of the enlistment. The RC soldiers were completely unprepared for mobilization. We had no clue as to what processes were required. It felt like everything had to be invented for family members, including new identification cards, health care cards, power of attorney documents, and other rights and benefits. No existing systems were established, nor were people trained to manage our situation. My only experience with mobilization was the established loading plan when we went to Camp McCoy for two weeks of annual training. This was a very new and different ball game.

Immediately, I contacted my commanding officer (CO). He asked me to report for duty when I returned to Wisconsin. Other unit members could not leave their civilian employment until the final reporting date. Within my unit, I was informed that I would be the transportation battalion movement officer for all our major equipment from Marshfield, Wis., to Fort Lewis, Wash. The following day, I attended a one-day training class in movement requirements. We were given a demonstration of rail loading procedures, a list of necessary loading materials, and a pre-approved purchase order to take to the local lumber yard. The following week, I directed and supervised the rail loading of our howitzers, wheeled vehicles, and transport containers of everything the unit had. I was surprised, yet relieved when all our equipment arrived safely in Fort Lewis with only one broken side view mirror.

An advanced party for each unit flew to Fort Lewis and the

main troop body moved by rail. On a crisp fall October day, our unit departed Marshfield. The somberness was felt in the cool, fall air. Everyone present was mindful that 20 years earlier, the Wisconsin 32nd Division had been called up during World War II. The Marshfield unit suffered heavy combat losses in the Pacific Theatre. Before leaving town, our unit marched in formation from the armory to the downtown train depot. The streets were lined with local citizens. There was no cheering or excitement. Everyone's faces were painted with grim and sad expressions. This was the first time that I said good-bye to my family, but it certainly was far from the last.

We boarded the train and waved out the windows as our journey west began. I felt like I was actually living in an old movie scene. And I was. As the train slowly headed west for the long ride to Washington State, we looked at each other, silently wondering what we were getting into.

It was no surprise that the no alcohol policy ordered for the troop trains was ignored. Until now, the culture of our National Guard unit had been more of a social experience than a professional one, as found with military enlistees. This difference began during my younger days at Camp Ripley. While we trained and could perform most tasks at least to the minimum standard, the training was not taken very seriously. Yes, we were soldiers, but the prevailing attitude was the military stuff was something we did in our spare time. National Guard soldiers all had something else they did for a living. None of us ever anticipated making the military our career.

Our RC units viewed AC Army people as those who were full-time soldiers. This important distinction lies at the heart of long-standing differences and animosity between the AC and RC. Still today, after many, many joint AC and RC deployments to Iraq and Afghanistan, I still hear rumbles about RC folks feeling like second-class citizens. There is a continuous opinion among the AC culture that RC units, especially combat arms, can never be fully mission-capable without extended periods of post-mobilization training. They should not be depended upon for rapid deployment. After decades of combined RC and AC service, I have lived and worked on all sides of this issue. My qualifications on this topic are significant. Realistically, I believe this controversy will always exist.

Along with my comrades, I consumed alcohol on the train trip. When all the alcohol supplies were exhausted, the train stopped at a small whistle stop somewhere in Montana. It was about midnight. We were informed the stop was for fifteen minutes. Across from an old wooden bridge was a very small town. We could see a lone lighted sign that flashed, “Bar.” The chief of firing battery, Wes, and I were selected to carry out an emergency resupply mission to the bar. Completing the daring mission would require a double-time run. Everyone agreed we were the most capable of accomplishing it.

Our uniforms consisted of the old-style Army green fatigues with boots and bloused trousers. We took off our jackets to increase our speed, which left us wearing white t-shirts. Our initial dash across the bridge was impressive. We had not considered how odd we would look bursting into a cowboy bar at midnight. At first sight, the people in the bar thought we were crackpot militants who had just come down from the mountains. Some of the bar folks seemed ready to defend themselves.

We assured them that we only wanted to buy a case of beer and would be gone as quickly as we arrived. After a few tense moments, the bartender jacked up the price and sold us the beer. As we escaped the bar and headed toward the bridge, the train began to chug, and the whistle blew. We bore down at full speed. I could hear the troops cheering as they leaned out of the train windows.

This would have been an appropriate time for the theme music from *Rocky* to bust onto the scene. But this was years before the movie was made. Halfway across the bridge, running as fast as possible, my fellow commando tripped and fell headfirst onto the bridge. It was a spectacular display of bravery. While he sacrificed most of the skin on his chest, he kept his arms extended and saved the case of bottled beer he was carrying from destruction.

Immediately, the cheers from the train turned to yells, "Hurry! Hurry!" The whistle blew, "Chew! Chew!" and the wheels picked up momentum. I did not want to be left behind.

As I grabbed the case of beer, I asked Wes if he could make it. He crawled to his feet and said, "Let's go!" As we closed in on the train, everyone shouted. Arms reached out to pull us aboard. We barely made it. It took weeks for Wes' chest to heal. The beer was gone in 30 minutes.

I have attended several reunions of our RC unit in recent years. Invariably, the train story always comes up. At our last gathering, we agreed to have a final toast to all our comrades when only a few of us remained. This feels like a better plan than a last-man club because none of us ever wanted to drink alone. I have secured a bottle of 1961 cognac for the special occasion when it happens.

## Fort Lewis

Draped in front of Mt. Ranier's impressive view, Fort Lewis is a special place. While it rains a lot, the weather is mild year-round. The entire Seattle/Tacoma area is a superb outdoor playground. It is one of the most desirable U.S. Army locations at which to be stationed.

We found it relatively easy to settle into the on-post facilities because they were so similar to those at Camp McCoy. All of the

32nd Division was placed on the north portion of the post, which included buildings identical to the wooden World War II buildings at Fort McCoy. We appreciated the same type of barracks, orderly rooms, supply rooms, mess halls, and other facilities. The first two weeks felt like just another annual training session. But then, reality set in. We wanted to know, "What do we do now?"

The 4th Infantry Division (ID) called Fort Lewis home. They were housed in permanent buildings on the main side of the fort. The 4th ID unit acted like our hosts, and this allowed our unit to partner with another battery that had a 155 mm howitzer. The 4th ID was helpful as we determined the lay of the land, local policies, and location of support facilities. Yet, their division had their hands full with their own mission and training. We knew they did not have much time to hold our hands.

One popular goodwill gesture was to trade standard issue mess hall supplies and condiments. The 4th ID had quite a few Hispanic soldiers. They took all the hot sauce we had. It did not feel like much of a sacrifice because none of our Wisconsin cheeseheads used hot sauce. In return, we gladly accepted their extra cheese.

One benefit of being an RC unit was that all of the soldiers in our battery had civilian jobs. These job skills added immense value to our basic day-to-day living and housekeeping functions. We had restaurant cooks, vehicle mechanics, construction workers, electrical and communication tradespeople. Our battery had the best barber in Marshfield. Our skills meant that we could easily take care of ourselves.

When training started, we were proficient and mission-capable at the battery level in short order. We could adequately oversee all of our basic living needs. However, staff functions at the battalion level and higher were a different story. It is difficult to effectively train staff personnel during peacetime. While I learned how to operate and maintain a howitzer when I was just 16 years old, I discovered that coordinating all field artillery duties is a whole different responsibility.

Eventually, the politics of our battery being called up became evident. Some soldiers' commitment to the cause deteriorated over time. We realized that there was no apparent intent to deploy us anywhere other than Fort Lewis. While the two National Guard divisions chosen for call-up were supposedly the best in the country, this was really an exercise to assess our combat readiness and deployability. Results were used to access the entire RC and influenced future structure decisions as to whether the RC could be mobilized.

It did not take long for our battery to realize the deck was stacked against us. No matter how hard we worked, we would never be declared combat-ready. I saw little evidence from the AC to help us improve. At the same time, we RC soldiers were not as good as we thought we were. Honestly, we spent too much time whining about the AC guys instead of being honest about our capabilities. Leadership was the missing ingredient. We simply did not have the necessary leadership that would take charge and get us ready. It was easier to just complain.

From the onset of our training, too much time was devoted to individual basic skill training. We were told that this was a prerequisite before we could begin collective training. Just a couple of months earlier, we had successfully fired all of our required artillery missions at Camp McCoy. We had completed individual weapons qualification firing as well. Yet, at Fort Lewis, we were treated like we had never trained in these areas before. It was very frustrating to go through this training again from the beginning.

Nine months after our National Guard battery was called up, the Army concluded they could not rely on the two called-up RC combat divisions to go to war. By then, this conclusion was no surprise to us.

## The Boland's Move West

Shortly after we arrived at Fort Lewis, we discovered there was no restriction on having our families join us. In fact, the Army would pay for moving and travel expenses. This was my first experience with permanent change of station (PCS) policies.

Soldiers were on their own to find housing in the area. Depending upon how long we stayed at a location, we could eventually become eligible for on-post family housing. Jackie and I waited until Christmas before moving the family to Washington. I flew home and drove the family back to Fort Lewis.

Our family went through many Army moves. Each time, we heard two persistent misunderstandings among our relatives and non-military friends about these moves. They believed military people had it made because the government paid for all housing and travel expenses. More than once, our family was told that we were living high on the hog at taxpayer expense while I served in the military.

Moving is one of the hardships of military life. Every time we moved, Jackie and I had to sort through two overriding worries: where would we live, and where would the kids go to school? Unfortunately, the burden of solving both these issues often falls upon the soldier's spouse because the new unit always wants the soldier to report for duty yesterday.

With more military families moving to Fort Lewis, it became difficult to find something suitable at an affordable price. I received a monthly housing allowance of $85.00 per month. Rental property owners near every military base are aware of these allowances and set rentals accordingly. I don't ever recall finding housing offered at less than the going rate. Most often, rates were higher. House hunting without a vehicle is challenging. Local agencies are more focused on selling a house than securing a rental. I have witnessed countless military families being taken advantage of as they try to find adequate family housing.

Realtors often begin discussions by inquiring if a soldier has used their GI home loan or not. If a soldier has not, the realtor encourages the family to purchase a new home with no down payment. This would often equate to lower monthly mortgages than monthly rent. While the concept sounds good, thousands of military families have used the home loan program available to them. So often, these families wind up with a house they have to leave, cannot sell, and cannot afford to keep. The result is countless foreclosures and loan guarantee eligibility benefits being forfeited. This situation always makes me feel ill as I think about the plight of military families just trying to get housing.

Shortly before Christmas, I found a decent three-bedroom, single-family home for $85.00 a month in the suburb of Lacy, Wash. Jackie had her first of many experiences dealing with the movers on her own. The new home worked well for us until the landlord contacted us and said they needed the house for a family emergency. Advised that it was not worthwhile to contest in court, we moved to a different house. The rent was $105.00 a month. Fairly new, we had genuinely nice neighbors. The house worked well for our children.

We stayed in this house until on-post family housing became available.

The Army pay scale is based on two variables: a soldier's rank and his or her years of service. When I was mobilized as part of the National Guard unit, I had over five years of total Army service. Thus, my basic compensation started at the over four-year level on the pay scale. The National Guard's required service time for an in-grade promotion requirement to First Lieutenant was either three years or 18 months of active-duty. I fell under a "whichever came first ruling." This meant I qualified for a promotion in June 1962, which was three years after my commissioning. For the rest of my military career, my early reserve component service always placed me higher on the pay scale, which became a helpful long-term benefit.

Our training at Fort Lewis progressed from individual training to battery level, then to battalion level, followed by a culminating division force-on-force maneuver exercise at the Yakima firing center. Each officer was required to perform several field artillery missions, including adjusting fire from an observation airplane. I had the opportunity to do this more than once. Each time, I couldn't help but think of the contrast in duties as a lieutenant flying a Cessna L-19 observation plane versus slogging through the mud with howitzers. Flying was a lot more fun. While I was not considering a field artillery career at this point, I realized that being trained as an Army pilot would provide me with a skill that I could later use in civilian life. I came to appreciate many things about the Army while at Fort Lewis. As a National Guard officer, it didn't seem likely I would stay very long on active-duty.

As I continued to seriously ponder about flying, another unforeseen development took place. We learned that the 32nd Division would be deactivated and returned to Wisconsin in August 1962. Another announcement indicated that the Army needed additional junior officers for a new buildup. We were told that any soldier in the 32nd Division with the rank of second lieutenant through captain who wanted to remain on active-duty could apply for Voluntary Indefinite status. If accepted, this route could provide a path for a National Guard soldier to potentially remain on active-duty for up to 20 years or until no longer needed. The Army requires a minimum of 17 years active-duty service to guarantee retention through 20 years of service, which qualifies a soldier for full retirement benefits from the U.S. military. Involuntary release before 17 years of active-duty service would mean no retirement benefits.

I saw this as a new bugle call opportunity for me. Was it a door that was opening up for me and meant to be? Maybe this was the call to duty that I had been destined for since my childhood. What should I do?

While it still felt unrealistic to consider the Army as a long-term commitment, I knew once I was classified as active-duty, I could apply for flight training. This would help me achieve an important short-term goal.

Unfortunately, I had no superior officer to turn to for mentoring and guidance. Most of the superior officers were anxious to go home. They did not understand why anyone would want to stay. Jackie and I labored over a decision. In my heart, I knew that I enjoyed serving in the Army more than teaching. I felt proud to serve our country and wear the uniform. After we tucked the boys into bed one night, we gave each other a hug. The next morning, I applied to remain on active-duty. Approximately fifty officers in the 32nd Division signed up, including several of my OCS classmates.

The Army must have needed us a lot more than I realized. The applications were quickly processed and approved. Within a few weeks, I was reassigned with orders to quickly report to the 1st Battalion, 30th Field Artillery, an active Army unit at Fort Lewis. My fellow officers in the 32nd Division hosted a farewell get-together for me. While they wished me well, a few of them thought I was wasting my potential in the Army. They thought I could accomplish much more outside of the military. Over the years, I have been asked what I would have pursued had I not remained in the Army. While I am not completely sure, I think I might have answered a call to serve as a basketball coach who built winning teams, just as Master Sgt. Mancini had taught me to do.

## My First Active Army Assignment

When I reported to the 1st Battalion, 30th Field Artillery, I was not surprised when their reaction was, "Oh no. One of those crybaby National Guard guys." I actually expected this welcome and was prepared for it. They did not realize that I chose this path. I felt more capable than the stereotyped image that preceded me. With the previous training at Fort Sill and a year of hard-earned experience behind me, I was confident that I could hold my own.

In my new battalion, I completed an internship as an assistant in the S-3 operations and training section. A tough Korean War combat veteran put me to the test. Under his watchful eyes, I gained his approval with various tasks and projects. Two months later, the battalion commander informed me that he wanted me to become the service battery commander (BC). This was a positive turn of events. The other BCs were captains. I had only been a first lieutenant for a few months. This was the first of three company-level commands with which I was entrusted. I worked hard to lead my commands well and be a team player.

There were very few regular Army officers in the battalion. Most officers were Army Reserve officers who were on indefinite active-duty service like me. However, I was the only National Guard officer. Most everyone thought it was odd that I was part of their battalion. In future years, the officer personnel management system went through a major overhaul regarding active-duty policies for reserve officers. In 1963, I was technically in the same boat as most other officers on active-duty.

While the 1st Battalion, 30th Field Artillery Battalion (1/30) was a separate Sixth Army Unit, it was under the operational control of the 4th Division Artillery Headquarters. When I reported for duty, I was processed through the Division Artillery supply personnel office. There, I met Second Lieutenant Powell, who, in subsequent years, became one of my best friends.

What this all means is that the 1/30 was a non-divisional unit controlled by a division. Our unit was organized differently than the divisional battalions. Our unit was assigned two L-19 observation aircraft and two pilots, called aviators, within the Army. Jackie and I became friends with the two pilots and their spouses. One of the pilots was a regular army (RA) officer. He was the only lieutenant I ever met who said he planned to stay in the Army for 30 years. I don't know if he quite made it. The last time I saw him, we were stationed together in Texas, and he had 25 years of service. Both of these pilots took me flying. I rode in the back seat of the L-19, also known as a Bird Dog. Eventually, they gave me some stick time, where I gained flying experience. Immediately, I was hooked on flying.

## The Army Family

Once I was a member of the 1/30, Jackie and I experienced a sense of closeness with the other officers, their spouses, and children we had never felt before. Most families from our previous unit remained in Wisconsin, so there were no social gatherings outside of our work hours. The reality sank in that we were extremely far from home. Our children were not able to see their grandparents on a regular basis. The adventure of being someplace new had worn off. Our little unit of four often felt very alone in a big world. The support of the other families became so important to Jackie and me. At the time, the public perception was that people in the military have a glamorous life. People outside of the military often believe that military families do not get homesick. Instead, families get used to moving because it happens so often.

This was not my family's experience.

Our family quickly learned the people and their families that a soldier served with became your family. A certain bonding takes place because soldiers do the same things together every day. Other soldiers are the people who know exactly what everyone is going through. Most of the folks back home do not understand what it means to serve in the military. Many are not particularly interested in knowing. Often, soldiers believe they only have each other and must depend on one another for mutual support through the good and the bad. As military families, we often did a lot of the same things folks back home did: picnics, birthday parties, Christmas parties, Easter egg hunts, and other special celebrations. As an active-duty officer, I built many special friendships with other soldiers that I served with. Some of these friendships lasted only during the current assignment. Some endured for a lifetime. As my years of service increased, my children found it only more difficult to move. They dreaded leaving friends behind. Unfortunately, they avoided becoming too close with their classmates because they knew they might not ever see each other again.

The Fort Lewis Officer's Open Mess was a happening place and the true center of social life on base. The 1/30 adults regularly had some fun partying together at the Officers' Club. Occasionally, there were command performance happy hour get-togethers. Everyone was expected to be there. I remember drink specials like pre-made 25-cent Martinis and Manhattans being placed on self-serve tables for quick access. Was this a good thing? No.

Our own unit had regular evening dinner dance parties where we learned to do the *Twist* with Chubby Checker blaring in the background. The officers embraced a great morale and spirit. We worked hard and worked well together during our training. This was an ideal environment for Jackie and me to transition into full-time Army life.

## Goodbye Fort Lewis

Convinced that I wanted to pursue the opportunity to become a pilot, I applied for the U.S. Army Officers Fixed Wing Aviator Course at Fort Rucker, Ala. When the commanding officer (CO) of the 1/30 learned of my application, he became irate and ordered me to report to him. Our conversation is one that I will never forget. He expressed his strong disappointment in me for being unappreciative of the opportunity he had given me to be a battery commander. The CO said if I forfeited the opportunity to be a commander, I was signing my own death warrant to ever have a successful artillery officer career. He believed no field artillery officer could ever succeed as an aviator. I was encouraged to heed his advice and change my mind.

Later, I realized how sincere and well-intended the CO was. He was trying to look out for me and give me sound advice. In retrospect, he was exactly right. I was not savvy enough to realize the long-term significance of the choice I made. Because I was still classified as National Guard, there was no likely path for me to achieve long-term active component service. I did not feel a career in field artillery could happen for me. I apologized to him and stayed with my decision to pursue flight school.

I was ecstatic when I learned I was accepted into the flight training course. I thought the stigma of being an active-duty

National Guard officer would eliminate me from consideration. I was ordered to report to Fort Rucker in April 1963. Because of my deemed "defection," I was exiled by the commander to the Yakima Firing Center as a liaison officer. This was one hundred miles away from Fort Lewis. Once again, Jackie was saddled with the moving details and arrangements.

On the night we left Fort Lewis, Jackie and I loaded the boys onto a platform bed we had arranged behind the back bench seat of our 1961 Ford two-door sedan. We drove all night so the boys would sleep. No seatbelts were required or installed. We did not even consider them as necessary. Our family was headed for the land of Dixie, unsure what this would mean. We were following this call I felt to become a pilot. Was this a viable choice? We would soon find out.

# Chapter 7

# Fort Rucker Flight School

Fort Rucker was and still is the official home of the U.S. Army Aviation School. Located in the southern Alabama area known as the Wiregrass, it is only ninety miles from the Gulf of Mexico.

When we arrived in Ozark, Ala., we stayed overnight in a motel. The next morning, we had breakfast at a nearby café. When our waitress asked us if we wanted grits, we had no idea what she was talking about. This was our introduction to Southern culture. We were surprised to find so many things different from what we knew in Wisconsin. Southern accents were something we had heard in movies, but we didn't realize that everyone still talked like this. After we finished breakfast and began to leave, the waitress called out in a traditional Southern saying, "Y'all come back, hear?"

We were no longer in the Midwest or Washington state.

As we drove from Ozark to Fort Rucker, we got our first glimpse of what the segregated South looked like. It was shocking to see these conditions in 1963. The housing and living conditions for many Black families looked unchanged from the Civil War period. Everything was segregated, including public restrooms. I had seen poor Black neighborhoods in Chicago, but nothing close to this. The civil rights movement was underway. We soon learned that the majority of people, black and white, did not believe any meaningful change would ever happen. The difference between the civilian community and the military community surprised me. By now, the military was integrated.

On a later assignment to Fort Rucker in 1970, our son Dan entered sixth grade. This was the first year a former Black high school became an integrated middle school. The middle school principal was highly respected throughout the community. With a PhD in education, the white folks called him the "Fessor." He ran a very tight ship, and there were no issues.

Once on base, I discovered another Army post with mostly World War II wooden buildings like Camp McCoy. All Army primary fixed-wing, or airplane, training was done there. A majority of rotary, or helicopter, training was performed at Fort Wolters in Texas. I was ordered to report to Fort Rucker in April, but my assigned class did not start until July. This was a 10-month instruction course with graduation slated for April of 1964. Until the course began, I was assigned to temporary duties at the Accident Investigation Board.

The first order of business was finding a place for our family to live. While we were eligible for on-post housing, there was a waiting list. We found a livable house nearby that was owned by a civilian flight instructor who didn't take advantage of short-term student renters. Neither Dan nor Dave had started school. Dan began kindergarten in the fall. A few months later, we were assigned to on-base family housing. It was a heavenly three-bedroom, air-conditioned duplex. We had wonderful neighbors. Life was good. Finally, we felt at home in the Army.

Our move allowed Dan to start kindergarten at the base elementary school. Our kids changed schools so many times over the years. It's a wonder they got through! Dan never attended the same school for more than one year; from kindergarten through his senior year of high school.

Two other 32nd Division officers were assigned to the same flight training class as was Lieutenant Powell from the 4th Division Artillery. Lieutenant Powell was a terrific singer who worked his way through college playing with a band. During flight school, Jackie and I enjoyed his glowing personality as well as his musical talent to the fullest. It was fitting that he had acquired the nickname "Shine."

When flight training began in July, the daily routine was challenging and intense. We rotated ground school and flying instruction each week from mornings to afternoons. The morning flying schedule meant reporting to the flight line at 6 a.m. These were long days with no let-up throughout the day. A number of students dropped out in the early stages. Once soldiers flew their solo flight, they usually finished the course. The course began with one hundred hours of training, called the primary phase. Phase two was 60 hours of tactical mission training. This phase was the most enjoyable as we practiced landing on various small fields or dirt roads. The L-19 was designed to land almost anywhere in Korea and could also take off within a very short distance. We learned how to do all of these maneuvers very well.

The final phase was 65 hours of instrument training in a De Havilland U-6 Beaver. We learned to fly relying completely on instruments with no visual reference outside the cockpit. This was the most difficult phase of the course. The navigational instruments and radios in the Beaver were primitive compared to today's planes. We were allowed no assistance from the co-pilot instructor. Truly, it was a challenge.

By the time we completed the entire course, our training level exceeded the hourly requirements for a Federal Aviation Administration (FAA) commercial license with a standard instrument rating. We were given the opportunity to take a written examination the day after graduation to receive an FAA license. Most of us did. My license is still valid, subject to meeting medical certificate requirements. Our instructors told us that we were the most proficient we would ever be on graduation day. They were right, based on mistakes I later made because I was rusty.

Flight school was not all work. We squeezed in some fun as well. Like all of American society, drinking and smoking were common habits in the Army in the 1960s. Most soldiers did both. The Officers' Club was the center of social life and sometimes a little crazy. Friday night's happy hour was a spirited event. Everyone attended, including students, instructors, staff, and civilian employees. A lot of singing took place with my friend Lieutenant Powell, "Shine" as the main attraction. Whenever I watch the scene in the original *Top Gun* movie with the soldiers singing a Righteous Brothers song in the bar, I think of our flight school days. We ALL thought we were *Top Guns*.

Shine was good enough that some students wanted to be part of his popularity. This led to the formation of a backup singing group. John Pfeiffer and I were classmates. We solicited George Volk from another class. He was an excellent electric guitar player and a brother to Fang Volk, an original member of *Paul Reviere and the Raiders*. Someone suggested naming this unlikely group the *Beavers,* after the aircraft we flew. This was also a takeoff on the immensely popular group the *Beatles*. The *Beavers* were born. We had fun imitating the *Beatles* songs. Quickly, things got out of hand. Everyone wanted us to perform at their events. It became too much, and our fame was short-lived. Soon, it was graduation day, and we all flew off into the wild blue yonder. Approximately a year earlier, the famous star Kris Kristofferson was a Fort Rucker student. Most folks probably do not know that he served as an Army pilot before his career in show business.

In April 1964, the night before our graduation, we had a big party at our living quarters. It felt like everyone showed up to hear the *Beavers* final performance. Our parents were also there, and they had a blast. George was also a skydiver. Before the evening was over, he had my dad in full jump gear, ready to jump the next morning. Thank goodness my dad did not jump.

As the years went by, I stayed in touch with the other members of the *Beavers*. Shine and his family remained close friends with our family. Now, 60 years after *Beaver* mania, only two of us are still living. John lives on the family farm in New York. I am in Wisconsin. Shine continued as an entertainer and became a popular singer in the Gulf Shores, Ala. area. There is an excellent YouTube video of him singing *Red Neck Riviera*. He was a wonderful friend and I miss him. One of the sad realities of growing old is the loss of good friends.

After graduation, our family remained at Fort Rucker, awaiting the arrival of our third son, Michael. My assignment orders were to Germany, but overseas travel was not allowed six weeks before or after birth. We left for Germany at the end of August when Mike was exactly six weeks old. We enjoyed our last days at Fort

Rucker with a few additional trips to the Panama City beaches, where Dan and Dave loved to play in the beautiful gulf waters.

Flight school had trained me to become a highly skilled aviator. As a commissioned officer, flying would not be my primary duty. The Army Aviation Branch did not exist yet, so aviators were members of other branches, such as the Infantry or Artillery. Flying was a secondary skill. Pilots were required to maintain proficiency, but our performance in branch assignments determined our future. It was very difficult to do both well, which is exactly what my CO at Fort Lewis tried to tell me. Aviation school gave us very little guidance on how to balance future assignments because their emphasis was on flying.

Most non-military people assumed I was in the Air Force when I said I was a pilot. Still today, clear lines remain between Army Aviation and the U.S. Air Force.

The Army Air Corps was the air arm of the U.S. military forces during World War II. Their theme song was the one I sang as a child. The song's last sentence says, "Nothing can stop the Army Air Corp." In 1947, the Army Air Corps became the U.S. Air Force and completely separated from the Army. From what I understand, the size and expanding technology of the U.S. air power had grown to such proportions that it was no longer manageable within the Army structure. In a more political sense, Air Force pilots were pilots, not infantry or artillery. Their roles are not all understood by their non-flying counterparts. They wanted to be on their own.

I have heard rumors that when the split became official, the remaining Army guys happily said good riddance to the pilots. A few Army pilots refused to transfer. They made the argument that the Army needed to maintain a small number of Piper Cub-type airplanes under the direct control of Army field commanders. They felt these were needed to provide observation, reconnaissance, and liaison missions, which led to more debate.

Another common rumor is that the founding father of the newly formed Air Force, General "Hap" Arnold, warned if the Army was allowed to keep any aircraft, someday they would have more planes than the Air Force. While I have not researched and confirmed these stories, I have heard these accounts from enough different sources to believe they are mostly true.

About 20 years later, General Arnold's alleged prophecy came true. The proliferation of Army helicopters during the Vietnam War changed the military aviation landscape. The Army started with a few small planes and the new name of Army Aviation. The Army kept the World War II pilot wing's design, while the Air Force created a different, larger wing and a new image. The Air Force also took the darn song with them. Now, the last sentence is, "Nothing can stop the U.S. Air Force." I have never been able to sing it this way.

World War II fighter pilot Richard Ira Bong's story demonstrates the inter-military service rivalry that sometimes happens. A farm boy from Poplar, WI, Bong, flew the P-38 in the Pacific Theatre. Credited with shooting down 40 Japanese aircraft, Bong is the all-time fighter ace in U.S. military history. He greatly exceeded the required five kills to become an ace. Bong was awarded the Congressional Medal of Honor by General MacArthur. Sadly, his legacy is mostly lost in the cobwebs of military history. The Air Force doesn't claim him because he was an Army pilot. Neither did the Army because he was an Air Corps pilot. I played a small part in helping establish a fitting memorial in his memory in Superior, Wis., close to Poplar. The memorial is a World War II Pacific-style Quonset hangar building that houses a restored P-38. Located along a main highway in northern Wisconsin, the memorial attracts many visitors. The book *Ace of Aces* shares his story.

## Goodbye Fort Rucker

Before leaving the States for Germany, our family crisscrossed the U.S. traveling with a newborn infant. So many military families have lived through this same challenge. I used some annual leave time to visit our kid's grandparents in Wisconsin and Chicago. They enjoyed seeing the new baby. I was surprised by everyone's response because Mike was a real screamer from day one. In every early snapshot we have of Mike, he is crying. Our port call for travel took us from McGuire Air Force Base to the Rhein-Main Airport near Frankfurt, Germany. Jackie and I had feelings of anxiety as we crossed the Atlantic Ocean. We wondered what life would be like in a foreign country.

I was assigned to the 212th Field Artillery Group in Hanau, Germany, as a senior staff aviation officer. My captain's promotion awaited my arrival. This was a much different artillery assignment than I had ever had, as well as my first aviation assignment. As part of the forward-deployed forces in Europe, the stakes were much higher than at Fort Lewis. Tensions remained high with the building of the Berlin Wall. There was no question. This was a 24/7 hardball mission.

I am not sure Jackie and I appreciated how good we had it in the Wiregrass area of southern Alabama. This lazy pace of life was quickly replaced by a vastly different season of life that was much more difficult. It is good that I did not know how challenging the bugle's call would quickly become.

*Author's note: In 2023, Fort Rucker was renamed Fort Novosel in honor of a Vietnam hero. For his gallantry action as a medical evacuation helicopter pilot, Chief Warrant Officer Mike Novosel was awarded the Congressional Medal of Honor. I did not know Mike, but his story is legendary among pilots.*

# Chapter 8

# Hanau, Germany

We arrived at Rhein-Main airport in Frankfurt, Germany, on Labor Day weekend 1964. Our sponsor picked us up on Tuesday and drove us to Hanau, which was less than an hour away. The effective date of my promotion to captain was August 26, just days before arriving at my new assignment.

Right away, I got off on the wrong foot. I had not coordinated my arrival closely with my new headquarters and instead relied on information provided by my old boss from the 1/30 Artillery in Fort Lewis. He was also in Hanau with his family. He was helpful and confirmed that we would initially stay in a hotel until we found a place to live because no on-post housing was available.

Unfortunately, he was in a different chain of command, assigned to the 3rd Armored Division Artillery, which was quite different from my assignment in the Field Artillery Group. I mistakenly assumed he shared the information I provided with my superiors. Regardless, the housing difficulties remained the same, no matter to whom we talked. We spent the first two weeks in a

German hotel where little English was spoken. We lacked a refrigerator in our room for Mike's formula. The whole situation put a strain on Jackie. Before the housing debacle was settled, we moved three times. She was ready to return to the states in short order. I felt very sorry for her.

The unit wanted me to begin my duties immediately. On the day of our arrival, the unit received a no-notice inspection from the Fifth Corps Inspector General's (IG) office. An IG inspection is a big deal no matter where a soldier is stationed. It was even more significant now because the aviation section for the group headquarters had completely failed the inspection. The assigned U-6 Beaver aircraft was grounded for multiple "red X" conditions, making the aircraft unsafe to fly. Within my first week of service in Germany and before my family had a place to live, I was standing at attention in front of the group commander, expected to explain the course of action I would use to correct this horrible situation. Aviation had suddenly become a major embarrassment to the command. They wanted action. Now.

As the ranking senior staff aviation officer, I was expected to correct the deficiencies even though I had inherited the issues. My predecessor was a career aviator captain who lived by his own rules. He left Hanau the Friday before I arrived and had been flying the commander and staff regularly in the Beaver aircraft that was now grounded. All aviator credibility in the headquarters had bottomed out. Welcome to Germany!

For the first two months I was at this base, the group commander arrived almost every morning at 8:00 a.m., expecting an aviation progress update. The Beaver aircraft's repairs were in the hands of a support maintenance unit. I had no influence on their work schedules and priorities. Neither did my group commander. I could only report the steps taken to improve in other areas.

Meanwhile, I began to orient myself to the local flying environment using the L-19 Bird Dog. Learning to cope with this geographical area and weather was more than a full-time challenge. Regularly, I flew to the Grafenwoehr Training Area, about a 90-minute flight in the Bird Dog, under marginal ceiling and visibility conditions for visual flight. The L-19 was not equipped for flight with only instrument conditions. There were instances when I had to climb into the clouds to a safe altitude and contact an air traffic controller for assistance to find a safe place to land. These landings were completed using only basic flight instruments and an Automatic Direction Finder (ADF) radio. We called these hairy situations because they truly were.

## A Major Setback

The lack of an adequately equipped instrument aircraft available soon led to another major problem. Regulations required that my standard instrument rating annual renewal must be completed by the end of my birthday month. Mine was due in November, and I arrived in September. For my renewal to be completed, I needed an appropriate aircraft for both practice and check rides and a qualified instrument examiner pilot to administer the oral and flight examination testing. I had access to neither. At the time, this was a common problem because there was no system like the Air Force to manage and oversee these kinds of requirements. In the Army, many aviators were on our own to make sure our training requirements were fulfilled. If an aviator was assigned to an aviation battalion where resources were readily available, the chain of command insured each aviator maintained his or her proficiency. In my case, the aviation chain of command consisted of me. The next higher aviation headquarters had an aviation staff officer who offered little help despite my urgent pleas.

The end of November rolled around, and I was unable to complete the required instrument rating renewal. According to regulations, I was automatically suspended from flying status and was referred to a flight evaluation board (FEB). They would determine if I should remain an Army aviator. My non-aviator-friendly headquarters was not impressed with my track record thus far. The person they depended on to restore respectability for their aviation section was now busted for failing to meet prescribed training requirements. I felt like I was at one of those rock-bottom moments. Yes, I had been in pickles before, but this one felt more substantial. I was on an overseas assignment without the resources to do my job. I felt like no one really cared if I succeeded.

A flight evaluation board comprised of other aviators was finally appointed. My case was reviewed, and I was called to appear before the board. After listening to my story and calling requested witnesses to confirm the facts, I was reinstated to flight status for 60 days so I could complete the instrument rating renewal. My future status was totally dependent upon meeting this requirement. If I did not complete these requirements, I could have been permanently grounded, a severe reality for me. Due to the visibility this matter had garnered, the next-level headquarters suddenly became energized and was willing to help provide the resources I needed to complete the training. After being in Germany for almost six months, my name was officially cleared, and I was fully certified to perform all missions. Our Beaver was finally out of maintenance and flyable. All other deficiencies had been corrected, and I could not wait to turn a new page with this assignment. I silently offered a prayer of "Amen" and convinced myself the worst was behind me.

During the spring and summer of 1965, I flew a lot. Gradually, this built confidence among my superiors. I flew regularly back and forth to Grafenwoehr, where most of the artillery firing training was performed. When I flew the Beaver, the flight was shorter than with the Bird Dog. My flight school classmate and friend, John Pfeiffer, flew similar missions for his unit. We saw each other regularly, especially at Grafenwoehr. John and I were based at the Fliegerhorst Kaserne, where the airfield was quite crowded with various Army aircraft. My memory says there may have been at least one of every fixed and rotary wing Army aircraft in Fliegerhorst's inventory. Six other Bird Dog aircraft were assigned to battalions subordinate to the 212th group. Altogether, there were seven Bird Dogs, one Beaver and one H-13 Sioux helicopter and helicopter pilot.

## German Life

Because of the Cold War, the U.S. Army was stationed literally everywhere in West Germany during the 1960s, with forces totaling around 250,000 personnel. It was essentially an occupation force, not entirely appreciated by the civilian population. It never occurred to me until later that World War II had ended only 20 years earlier. Local Germans still had very strong feelings about the heavy bombing of the area where we lived; bombing they felt was unnecessary. The general consensus was this bombing came from British forces in retaliation for the German bombing of England. This happened towards the end of the war when it was essentially over.

As a U.S. military member, I would rate the German's feelings toward us as lukewarm at best. It was difficult for American soldiers to adjust and fit into the local culture. Many Germans were friendly, but most remained distant. They wanted us to speak the German language in the community. Fortunately, I studied German in high school and college and stumbled along with my language skills fairly well. It was common for Americans who only spoke English to be ignored by employees at local business establishments.

Based on clothing, hair styles, general appearance and manner, it was very easy to distinguish the Germans from the Americans while walking down the street. When I returned to Germany 20 years later, this all had changed. All Germans, especially women, looked more like Americans.

When Jackie and I arrived in Germany in 1964, the monetary exchange rate was four Deutsch Marks per one U.S. dollar. This very favorable exchange meant purchasing German goods and services was a complete bargain. The excellent German food, beer, and wine were very affordable. This changed significantly during this assignment. By the time we left in 1966, the exchange was more like 2.5 Deutsch marks per U.S. dollar. When I visited Germany 50 years after my first experience, the food, beer, and wine were still wonderful but a little less affordable. It is tricky to draw a complete exchange comparison because the Euro dollar is used today. I would still rate eating in a German restaurant as a good value. Today, everyone really looks the same. It's nearly impossible to tell an American from a German.

## Flight Pay

As an aviator, I already felt it was difficult for me to fit in and gain acceptance as a team member in an artillery unit. Flight pay made this even more challenging. Many non-aviators brought this to my attention regularly because it was such a sore spot with them. As a captain with over eight years of total service, my base pay was $565 per month. My flight pay was an additional $185 per month. Totaled together, my total pay was close to that of a lieutenant colonel (LTC). Ironically, my artillery group executive officer (XO) was an LTC. He also endorsed my written performance evaluations.

Not long after I arrived in Germany, he called me into his office, trying to deal with the aftermath of the inspection debacle. In no uncertain terms, he told me that he had no use for Army aviators. He emphasized this by stating that he would never be a passenger in an aircraft flown by an Army aviator. While he felt aviators should be paid flight pay, he did not understand why aviators received base pay. I was stunned by his comments, especially when I realized that he was serious. Technically, flight pay was defined as hazardous duty pay. Unfortunately, the XO didn't appreciate what this meant. Before very long, both my immediate boss, an operations officer (S-3), and the group CO would have direct experiences that made them more appreciative of the hazards aviators dealt with.

Non-aviators also complained that aviators cried wolf about bad weather when they didn't want to fly a mission. Aviators could never quite convince non-flying personnel just how difficult and challenging the weather was in the area where we were stationed. Every day, aviators dealt with relentless weather conditions. Grafenwoehr was often socked in during the morning hours and would hopefully clear as the day progressed. Aviators were required to file a flight plan for every mission we flew, accompanied by an official and written weather briefing. Most of the time, it was impossible for a weather forecaster to guarantee these visual conditions from Hanau to Grafenwoehr. Military aviators were dependent upon an arrangement that allowed us to fly into local flying airspace, a 50-mile radius from an airfield. Weather forecasters would give us a forecast for a local flight plan of 700-foot ceilings in the valleys and one mile of visibility.

The edge of the local flying area to the east was near Wurzburg. We would wind our way along the Main River that ran from Hanau to Wurzburg. There were a couple of tight turns, which made even a 700-foot ceiling and one mile of visibility very dicey. Sometimes, the weather forced us to turn around and return to the base. Making these flights became an all-day hopscotch process, getting from one safe location to the next, which required us to file multiple flight plan changes en route. We avoided performing these flight plan changes when passengers were on board. Most often, aviators crept through this process when it was necessary to accomplish some other business or transport something.

One evening, my S-3 major called me at my quarters. Major Ed Queeny needed to be transported to Grafenwoehr immediately. I informed him that the weather conditions would not allow us to fly. He did not want to hear any more weather BS and expected me to meet him at the airfield ASAP. I called the weather forecaster. The best they could give me was a local clearance, and confirmed there was no way we could get to Grafenwoehr. It was raining, and visibility was minimal at best. As ordered, I met Major Queeny, and we took off into the darkness. I knew how far I could go and remain safe, but I also knew it would not be pretty. I was convinced the only way he would believe me was for him to witness it firsthand.

We didn't get very far, and it became nasty. It was raining so hard that it was impossible to see. The wind kicked up and bounced the aircraft. He told me that he thought we should head back; I knew where we were. We had sufficient altitude for terrain clearance but nothing more. I continued to maintain our course towards our destination and said nothing. In a few minutes, he again urged me to turn around. Suddenly, he could wait until morning and go by ground transportation if necessary. I told him I was willing to continue to fly. Finally, he ordered me to return to Hanau.

Back on the ground, he seemed like a changed person. Major Queeny thanked me for trying so hard to accomplish the mission. He never realized what it was like to fly in those conditions. He appreciated how I had remained calm and professional. From that moment on, our relationship completely changed, and in time, we became friends.

Eventually, Major Queeny and his wife, Ann, grew very close to Jackie and me, as well as our children. For many years, we shared wonderful times together. At the end of his career, he retired as a lieutenant colonel and lived in Oklahoma when I was assigned to Fort Sill. In 1978, when our son David was a high school senior, he played on a top-ranked football team that marched to the semifinals in the Oklahoma state championship. Ann and Ed traveled to see Dave play as if he was their own son. Because the Army brought us together, we became family.

## On A Wing and a Prayer

On another occasion, I had a mission to take the group CO and command sergeant major to Grafenwoehr. Most of the staff, including my boss, were there for a training exercise. They planned to present an update briefing to the CO upon our arrival. We planned to depart Hanau late afternoon and arrived at Grafenwoehr around 6 p.m. The weather forecast was tight, but we had a flight plan clearance to go.

Often, Grafenwoehr was socked in at dusk and would become impossible to fly into. We departed with two Bird Dogs on a joint flight plan. As we approached Grafenwoehr, the weather deteriorated. Grafenwoehr was only a five-minute flight from the Czech border, which could not be overflown. A buffer area known as the Air Defense Identification Zone (ADIZ) was adjacent to the border. Rules required aircraft to penetrate the ADIZ within three minutes of the flight plan estimate. Radio jamming from the hostile side of the border was common. This was a way to distract aircraft and cause border overflight so the fire could be engaged. It was a high-stress area with no margin for error.

As we were about to penetrate the ADIZ, we lost visual contact with the ground. I was in radio contact with the Grafenwoehr control tower and reported ADIZ penetration. The tower operator confirmed my position report but cautioned that the fog was closing in rapidly at the airfield. We already knew that the Grafenwoehr ADF was not operational. An instrument approach to landing there was not possible. I looked down and saw a brief opening in the clouds. I could see Vilseck Airfield, our last reporting point. After I reported this, the tower operator more urgently conveyed the worsening visibility in the field. Just then, the other aircraft, my wing ship, informed me that he was completely "in the soup" and had also lost radio contact with the airfield. He only had contact with me via FM radio.

I directed him to climb 500 feet above me, slow his airspeed, and maintain his current heading. I immediately radioed the tower and informed them of our situation. Then, I requested a practice ground-controlled approach (GCA). We were in violation of Army flight rules, but we were in deep trouble. I just wanted to get both aircraft on the ground. I felt this was our only viable option.

The tower operator informed me that the radar operator had just been shut down, and he was headed to the parking lot. The tower operator yelled to the radar operator and asked him to run back to the radar and fire it up. All of this happened within a few minutes. We also needed to turn back towards the inbound heading for the field. About then, I realized the colonel sitting in my back seat had not said a word during all of this. Thankfully, he let me deal with the situation without raising any questions.

The GCA operator quickly picked us up because there were no other blips on his radar screen. He talked me through the landing. I had to relay all his instructions to the other aircraft. It was hectic throughout the landing. I confirmed each instruction on my UHF radio, then quickly switched to the FM radio and relayed this to the other pilot. I went back and forth between the two radios multiple times. Finally, the GCA operator said, "On course, on glide slope, a quarter mile from touchdown." I finally saw the runway lights and throttled back the aircraft for landing.

After touching down, I cleared the runway as soon as possible so the other aircraft could land. As I did this, I looked back, saw his landing lights and watched him touch down. By the time we both taxied to the parking ramp, the airfield was socked in with completely zero visibility.

The S-3 major and other staff were waiting for us. As we exited the aircraft, they said what we expected, “Wow! We weren’t sure you were going to make it!”

I resisted saying, “Neither did we.”

The staff told the colonel they were ready for the briefing with the rest of the staff standing by. The colonel told them, “The hell with the briefing. I’m going to the Officers’ Club and buying these pilots a drink!”

And he did. After the colonel gave a full account to everyone present about the amazing job his pilots had done to complete the mission, I didn’t hear any more sniping about flight pay.

The next morning, I quickly learned how success can evaporate. When I arrived at the airfield, I was informed that the airfield commander wanted to see me immediately. He was a lieutenant colonel aviator who was intolerant of any infractions of flight procedures at his airfield. After properly reporting to him and saluting, I received a royal chewing out. He ripped me apart for violating visual flight procedures and making an instrument approach with an aircraft not equipped for instrument flight. He gave me no chance to explain and said he was tempted to file an official flight violation. He also recommended that I go before the Flight Evaluation Board (FEB). Quickly, I decided this was not a good time to share that I had recently undergone an FEB. I knew another infraction would likely end my flying career. Obviously, it was a time to beg for forgiveness and hope that he would let me off the hook.

He made it clear that he preferred not to see me again on his airfield. If I did return, there would be no second chances. I assured him that there would be no further cases of concern. I returned to this airfield often and was ever mindful of my fragile status.

## Tailwheel Ray

The L-19 Bird Dog was a tricky airplane to fly. It required a lot of practice to remain proficient in all of the required maneuvers and procedures. The Fort Rucker flight instructors were accurate when they told graduates that we were the best pilots we would ever be when we completed flight training. This was the case with short takeoff and landing (STOL) procedures. In flight school, we practiced these every day. From the time I arrived in Germany, all my landings and takeoffs were on lengthy paved runways. Very few field training exercises required us to land on unimproved tactical settings.

I was informed that my group commander planned to visit another headquarters during a field exercise. I would be the pilot flying him in the next few days. Wanting to make sure things went smoothly when I took the CO, I checked things out ahead of time. The helicopter pilot rode along to experience short takeoff and landings in the Bird Dog. He had some doubts about my abilities, but I assured him that he would be impressed. We did a map reconnaissance of the location and headed out.

When we arrived at the area, I circled, looking for the landing field. We had been told the area was quite small, but the unit had several Bird Dogs and used the landing area. After a few minutes of circling, we could not find the field. Suddenly, my passenger saw a Bird Dog passing under us at a low altitude. We watched it complete a landing approach into an extremely small field. As I descended to take a closer look, the helicopter pilot asked if I was going to try to land in the same spot. I was a bit cocky when I said, "Sure. If they can land in there, so can I." He preferred that we return to our base before I attempted the landing. I told him, "Don't worry. I'm trained for this."

There was only one way in and out of the field, and one end had a road with a fence crossing. The other end had tall trees. This left little room for a landing roll after touchdown. I set up a good approach with a steep descent angle. The procedure required using full flaps and lots of power to maintain airspeed. It looked good all the way down, and I was confident that we would land on our first attempt.

As we reached the short final approach point, I made a mistake and reduced power too soon. The aircraft stalled at about 50 feet above the ground. We hit extremely hard and immediately bounced back up into the air. I realized the trees were too close to salvage the landing. Instead, I applied full power to climb out of the landing. As we began to climb, I felt a strong vibration in the aircraft's rear. The tailwheel was broken. My passenger could barely wait to ask me what I planned to do now. As I answered, I didn't think about the folks on the ground where we had attempted to land who might be wondering about us.

Our only option was to return to our base and attempt to land without further damage or possible injury. In-flight training, we were drilled daily on making textbook three-point landings. This required getting the tailwheel firmly on the ground at touchdown, which would establish positive directional control during the landing roll. Failure to do this often resulted in a ground loop and significant damage to the aircraft.

The Bird Dog was notorious for being a ground loop waiting to happen. To a lesser degree, we were also trained to perform main gear-only landings under certain conditions. These were not intended for broken tailwheel scenarios. This landing required the use of flaps and power to remain above stall speed with a shallow runway approach rather than a steep one. In other words, this is an extended, flat, final approach to drive the aircraft onto the runaway. After the touchdown, pilots used the proper critical technique to avoid letting the propeller strike the runway. It was another tricky procedure that I had not attempted since flight school.

My passenger had no reason to trust my skill, but he was so anxious to get back on the ground. He left me alone to concentrate on the approach. The tailwheel was still causing significant vibration but did not interfere with the flight controls. I didn't want to declare an emergency landing to the tower operator, as it would attract lots of attention I did not want, including another investigation. I informed the tower operator that I would be making a practice main gear landing with a full stop because we had minor tailwheel difficulties. I might exit the runway onto the grassy area short of the taxi way. I was cleared to land but asked if I wanted emergency response with fire trucks, etc. I assured him this was not necessary because our situation was routine. As I took a deep breath, two quick thoughts entered my mind. "Why does this shit happen to me?" and "It's your own damn fault!"

Motivated and inspired by my previous screw-up, I made the most perfect main gear landing. It was a thing of beauty. I controlled our touchdown speed properly to limit forward roll and get the aircraft off the runway before the tail touched the ground. Before shutting down the aircraft, I informed the tower operator that I would have maintenance personnel check everything out before we repositioned on the parking ramp. As we exited the aircraft, the helicopter pilot confirmed that I was correct in what I said at the outset. He was really "impressed."

While this story could end here, it was far from over. The rest of the story is also true but tends to be far less believable.

## Practice Makes Perfect - Almost

As we settled into the flight operations building, everyone wanted to know what happened. My buddy and I had slightly different versions. The bottom line was we had a broken tailwheel that was repairable, and damage was minimal. The more I thought about what happened, the more I became upset with myself. I knew that I must accomplish the colonel's mission without difficulty or incident.

I kept staring out the window at the other Bird Dogs on the ramp. I wanted another try, but sensing my thoughts, the helicopter pilot said he hoped I wasn't going to do what he was thinking I might. When I told the crew chief to ready another aircraft for flight, the helicopter pilot told me to "count me out" and left the building. The other pilots departed. It was just me and the crew chief, dutifully readying the plane.

Within two hours of the first episode, I headed back to the field site by myself. On the way, I rehearsed the landing in my mind a dozen times. I knew exactly what I had done wrong and how to correct it. I set up the landing approach and talked to myself all the way to touchdown. It was perfect. I came to a stop, did a 180-degree turn and taxied back to the approach end. Now, I could see other aircraft, vehicles, and people tucked into the tree line.

One of them guided me with arm signals to a parking spot. I shut the engine down and got out. I almost received a hero's welcome. Someone said, "Wow, that was a terrific landing. You're the first one to make it with one approach. It took several tries for all of us before we could get it right." I agreed that it was definitely a snug spot, but still cocky, "It wasn't that bad."

Then someone said, "Some crazy SOB tried to land there a couple of hours ago. Damn near crashed and burned in the process. We don't know what happened to him afterward. But his tailwheel was spinning and flopping all over the place when he flew away."

"Wow," I said. I didn't have the gumption to admit that I was the SOB they were talking about.

They asked if I heard about any aircraft emergency or crash when I was in the air. I hadn't, but I would keep my eyes and ears open on the way back. After a few more minutes of chatting, I informed the group that I would be returning soon with my CO when he visited their exercise headquarters. I looked forward to seeing them again.

With that, we rolled my plane into the open area. I got in and started it up. Before I taxied away, I gave them the traditional thumbs-up gesture through the window, just like in the movies. As I taxied toward the other end of the field, I realized that I needed to get as close to the trees as possible. I needed every inch of space for takeoff. During our group discussion, the guys told me there were only 600 feet of usable runway.

When I got as close as I thought I could, I planted the left brake, applied power, and started a 180-degree turn to the left. At the 90-degree point, I realized the rightwing tip was touching a tree. I could not complete the turn. To back up, I would have to shut down the engine, get out, and move the plane rearward by hand.

As I got out, I did not set the parking brake so I could move the wheels. I did not realize there was a significant slope to the ground at the plane's rear before I pulled on the ground handle located near the tail. When I tugged the handle, the aircraft began to roll quickly out of control. I was nearly struck as it sped by me, headed downhill backward. I instinctively gave chase, but it was moving much faster than I could run. Before I ran very far, I went head over heels and turned my ankle. As I got on my knees, the plane was still rolling backward in a perfectly straight line. I rose to my feet and limped along the plane's path.

Then, I noticed the soldiers were speeding down the hill with a truck, several of them hanging off the sides. It looked like something out of one of the old *Keystone Kops* movies. Gradually, the plane slowed down, nearly to a stop. It disappeared behind a slight rise in the ground.

I continued ahead and watched the vehicle stop, and the soldiers get off. They stood along the rise of the ground, looking down at the plane. I still could not see it. They were shaking their heads and waving their arms. I assumed the aircraft was completely destroyed. When I finally reached where they were standing, I shared their disbelief that the plane was completely intact and looked like it was deliberately parked where it sat. When it first began to roll backward, the tailwheel inverted from its partly turned position, locked up and could not turn. The attached leaf spring continued to bend backward and began to dig into the ground. This is what slowed the plane to its eventual stop. The bottom line? I had my second busted tailwheel of the afternoon and a severely damaged ego.

As we all continued to stare at this amazing sight, one of the guys said something like, "This has been a helluva day!" Then, someone asked me what I was going to do. After looking closely at the tailwheel area, I asked if they had a spare tailwheel. One of the crew chiefs said he did. I asked if they would loan it to me and if they could help install it. They were agreeable but could not guarantee how long it would hold up due to other damage and weakness to the tailwheel assembly. I accepted full responsibility by signing off on the necessary entry in the logbook authorizing a one-time flight.

The conversation took another turn when they asked what I intended to do once the tailwheel was replaced. My response raised a strong question of my sanity by the way they reacted. I intended to start the engine and determine whether everything was still working. If so, I planned to taxi the aircraft back up the hill, along the same path it had come down, about 200 feet. When I made it back to where I started, I planned to take off and return to my base. I think the same guy said, "This IS a helluva day. First, one Bird Dog almost crashed. Now, we have another one about to crash."

They raised major doubts about my plan, with their main concern that they would end up with a major accident on their airstrip. I showed appreciation for their concerns, but unless someone had a better idea, I was going to try. I followed through with my plan, and it worked.

Airborne, I headed for home base. I felt no vibrations and concluded the tailwheel was still attached. I made a normal approach at Hanau with as soft a touchdown as possible and then a normal taxi to the parking ramp. By this time, my troops had reassembled. They anxiously awaited my return or possibly the news that I would not be returning. The million-dollar question was whether I had landed in the pea patch or aborted the mission. I told them that I landed with one approach, and it had been a piece of cake.

The pilots were duly impressed. I mentioned that the aircraft seemed to have a tailwheel problem. It should be checked out, and I did not share the rest of the story with them or anyone else for a long time. I wondered if anyone would believe my storied tale.

Eventually, the guys at the small field found out that I had flown both aircraft. Two years later, I overheard a conversation in a Vietnam bar amongst a group of Bird Dog pilots. One guy told a story about a crazy SOB who crashed two Bird Dogs in one day during a training exercise in Germany. They were all laughing their butts off. Like any good war story, the details had grown even more ridiculous. I was tempted to correct the details but wisely resisted. This decision indicated my judgment might be improving.

In recent years, I met a guy in Wisconsin who owned a Bird Dog. There are quite a few still able to be flown. We became friends. When he found out about my military pilot experience, he insisted I fly his Bird Dog. Although it had been 35 years since I flew an L-19, I accepted his offer. Amazingly, it felt like I had never stopped flying it. He and his Warbirds flying group adopted me and made me an honorary chief pilot of his plane. After sharing my old German story with him, he had the words "Tailwheel Ray" painted on the tail. I flew his newly painted plane all over Wisconsin and gave rides to my sons and grandchildren; it was a thrill for all of us.

## Defense Counsel

In the summer and fall of 1965, I logged lots of flying hours, which helped build credibility with my superiors. I was also assigned other additional duties and projects, where I demonstrated my ability to perform tasks other than flying. I worked hard to do everything I was given in the best possible manner. This started to reap dividends with my bosses. It never occurred to me that I might do a job too well and end up in the doghouse. Soon, I first experienced the mantra that if something could go wrong, it often will.

In the 1960s, Army junior officer professional development included requirements to perform special duty as counsel in military court martial proceedings. We were assigned on standing orders for six-month periods as either trial or defense counsel for cases that were brought to trial. On one occasion, I was appointed to defend a soldier charged with misappropriation of a privately owned vehicle. On the date in question of the alleged crime, the defendant had committed a series of offenses, including speeding while driving drunk without a valid driver's license using a stolen vehicle.

Conviction for the lesser offenses would have been an easy no-brainer. If convicted, the stolen automobile charge carried the most severe punishment. This is the route chosen by the convening authority and group commander.

Preparation for a court martial required significant study of the Uniformed Code of Military Justice (UCMJ), the military justice system guide. Knowledge and experience gained from knowing the guide have long-term value to every officer, especially if one is later assigned to command duty.

From my own investigation of the case, I learned the vehicle in question was abandoned. No one knew who the vehicle's last owner was. It had been in the unit parking area for a long time, handed down from one soldier to another. In effect, it had become community property for anyone to use. Why this situation existed without the chain of command's knowledge was probably a worse violation than the charged offense. But this was not the issue on trial.

It became relatively easy for me to establish reasonable doubt that the vehicle was stolen by the accused, and he was acquitted. Unfortunately, this did not sit well with the convening authority. They certainly made it known to me that I had screwed up. I realized that no good deed goes unpunished.

Eventually, the system was changed. The practice of unit officers performing counsel duty was eliminated, which helped reduce command influence in punitive actions.

## Time To Move On

As I gradually became more adjusted and successful in my first flying assignment, the U.S. involvement in the Vietnam situation was escalating. While very few of my flight school classmates went directly to Vietnam, one went down in a Bird Dog and was missing in action (MIA).

President Johnson announced a full-scale commitment of U.S. forces to Vietnam, starting with the Army's first airmobile division, the First Cavalry. The demand for helicopter pilots grew so rapidly that a special rotary wing transition training program was set up in Germany. All fixed-wing pilots in Germany would attend this course and be sent to Vietnam. Most of the fixed-wing guys were not enthusiastic about flying helicopters, especially in Vietnam. We mistakenly believed that under emergency conditions, an airplane was safer than a helicopter. I was among those who wanted very much to remain a fixed-wing pilot if I had to go to Vietnam. I finally felt experienced enough in airplanes to be combat-ready.

As I was pondering the situation, I had an opportunity to meet with an assignments representative, a major from the Field Artillery Branch, who was visiting units in Germany. We discussed my record and performance evaluations. For the first time while serving in the military, I benefited from meaningful counsel and received informed advice on how I might improve my situation.

Up to this point, my written performance Officer Efficiency Report (OER) evaluations were something about which I was very naïve and uninformed. My National Guard officer training and assignments had not educated me on the great importance of these evaluations. While I had received very low scores from my National Guard supervisors, my scores greatly improved during my brief time with the 30th Artillery Battalion. After I jumped ship and went to flight school, I received similarly improved scores.

Based on my recent written performance OER evaluations, the major could find me a pathway for a long-term opportunity. I had convinced the Army I was ready to be a battery commander. Yet, he informed me I was well behind my peers and had my work cut out to ever catch up. This, along with my National Guard status, did not bode well for me.

I needed to continue my recent improvements and establish a steady upward trend in my performance rating scores. It was important that I maintain consistency without any downturns on my record. If I wanted to be considered for future advancement, I could not afford to stumble. There were enough positives in my record to believe I could succeed, provided that I worked hard every day and gave everything my best, no matter what circumstances or personalities I encountered.

Frankly, I wasn't sure if the whole conversation was important because I had no intention of remaining in the Army long-term. My goal was to become a pilot and complete the required three years of service and resign. I planned to become a commercial pilot, like so many other military pilots. As I listened to the major, I began to realize that I needed to grow up and not burn any more bridges. Going forward, I discovered that the OER was a frequently discussed subject among fellow officers.

As we talked about fixed-wing versus rotary-wing aircraft and Vietnam, he made a helpful suggestion. From a field artillery standpoint, what I flew didn't matter as much as how well I flew. He encouraged me to demonstrate a strong military commitment as an officer, volunteer for Vietnam and not wait for the upcoming helicopter training. If I did this, the Army would honor my request and send me to Vietnam as a fixed-wing pilot.

This option made sense to me. I would be going to Vietnam either way. I was still under obligation to serve because of my flight training requirement. If I volunteered now, I would go sooner on a fixed-wing assignment. While the offer was tempting, I wasn't sure my wife would understand if I volunteered to go to war. His last comment truly captured my attention. He told me this was my career. Only I could make the decisions and take the necessary actions to succeed. Serving in the Army carried special responsibilities. Some of these decisions will always be difficult for spouses to understand or appreciate.

I went home and had a long talk with Jackie. She was very unhappy living in Germany and was anxious to return to the States. She didn't want me to volunteer unless I was absolutely convinced this was the best thing for me to do. As I looked at our three little sons, it became even more difficult to resolve my decision. We talked about it over the next several days, and then I made the decision to volunteer for Vietnam.

This was a very difficult decision for me, Jackie and my family. While it was made over 50 years ago, the stress of trying to sort through this choice remains very close to my heart today. I do not know how other soldiers feel, but I have great empathy for today's volunteer soldiers who have withstood multiple deployments into war zone duty. The way military families are stretched nearly to a breaking point is a reality of their lifestyle, yet so often unknown to others.

In April 1966, we left Germany and returned to the United States on a troop transport ship named the Darby. Soon, this same ship transported many soldiers to Vietnam. We rented a house in Tucson, Ariz. Jackie's sister, Pat, lived there with her husband, who was training to become a doctor. Knowing Jackie would be near family while I would be gone for a year was very important. This became a great blessing, as our families spent a lot of time together in the upcoming months.

I used all the accumulated leave time I had to settle our family and prepare for war. The weather was amazingly hot in Tucson. People often say it is not as intense because it is a dry heat. I never bought into this idea.

A couple of days before heading to Vietnam, I received all the required immunizations at the nearby Davis Monthan Air Force base. The tearful hugs and kisses with Jackie and the boys from the day I left Tucson will always be etched into my heart and soul. As I slowly walked towards an airplane that would take me to California, I fought back my own tears, standing on the stairs as I turned to wave. For a brief moment, I realized this same scene was taking place with hundreds of families all over America.

# Chapter 9

# Good Morning, Vietnam

Next stop: Vietnam. Unfortunately, a serious incident occurred before my feet touched Vietnam soil.

The flight from California to Vietnam was far more unpleasant than expected. Waiting until the last minute to get immunized was a bad idea. Shortly after takeoff, I realized my left arm was sore. Soon, it became inflamed and started to swell. By the time we arrived at our first stop in Hawaii, my arm was terribly swollen at the injection site. The swelling gradually moved down my arm, which now looked like a bowling pin.

We were flying on a contracted commercial airline. There were very few Army personnel and no medical help available at the airport in Hawaii. I wore the prescribed short-sleeved khaki uniform. The swelling was so intense that the sleeve's tightness was painful. The only medical assistance I received was a pair of scissors used to cut the sleeve vertically towards the shoulder to relieve pressure and a couple of aspirin. Despite my pleas, no one was authorized to remove me from the plane for treatment. I was in a "suck it up, buttercup" situation. Not convinced I could last another eight hours in the air, I had no other option.

By the time we landed in the Philippines, I was much worse. Again, I found no sympathy or authority to help. Their mantra was, "It's only a few more hours to Saigon." I don't remember landing in Saigon. My first memory of Vietnam is lying in a hospital bed, receiving major doses of antibiotics. The doctor confirmed I was very sick.

Good morning, Vietnam!

I was in the hospital for several days before I started to feel better. Meanwhile, I became acquainted with my new environment of heat, humidity, mosquitoes, and the ever-present distinctive odor of wet, burning wood. Quickly, I discovered the importance of mosquito netting.

The hours dragged by slowly. As my health improved, I wondered where my assignment would be. Everyone in Vietnam counted the days until they could return to the U.S. When someone asked, "How are you?" the answer would be something like, "236 days and a wakeup." I was pleasantly surprised when I discovered that my first days in the hospital counted toward my 365 days in Vietnam.

When released from the hospital, I was sent to receive my assignment as well as transportation to my new unit. While I had succeeded in going to Vietnam as an airplane pilot rather than as a helicopter pilot, possible assignments were limited. I had requested training for the Caribou Cargo aircraft but had been denied. I expected to be assigned to an L-19 Bird Dog unit. The only other aircraft I was qualified to pilot was the Beaver. There were very few of those in Vietnam. By 1966, there were several Bird Dog units in Vietnam. The main question was what area of the country I would be flying in.

The assignment notification procedure was simple. Every morning, listings were posted on the replacement company bulletin board. Soldiers looked for their names, and I was assigned to the 17th Aviation Group in Nha Trang. This was a major headquarters for aviation units in the Second Corp operations area and included the region known as the Central Highlands. A quick check of rumor intelligence among fellow aviators at the replacement company suggested my ultimate destination might be the 219th Bird Dog Company, also known as the Headhunters, at Pleiku.

The next day, I was on my way to Nha Trang. When I arrived, I was directed to the 14th Combat Aviation Battalion located at Lane Army Heliport near Qui Nhon. Nha Trang and Qui Nhon are located on the central coast of South Vietnam. Nha Trang was a favorite seaside resort area during the French colonial period.

After a short hop on an Air Force C-130 Hercules, I was on the ground in Qui Nhon and transported by ground to the 14th Battalion headquarters. The C-130 was a workhorse transporting people and cargo all over South Vietnam. While waiting for my ride, I ran into Major Billy Bartle. He was the executive officer of the 18th Aviation Company, which flew the U-1A Otter aircraft. The Otter is the big brother to the Beaver. De Havilland Company built both aircraft for bush pilot operations in Alaska and Canada.

Major Bartle said pilots with experience flying the Beaver were needed to fly the Otter, which had a similar design to the Beaver but was much larger. He asked me to request an assignment to the 18th when I interviewed with the 14th Battalion. I followed his advice. My request was granted.

Because the 18th Aviation headquarters was in Nha Trang, I got another C-130 flight back there. Already, I looked forward to flying the big bad Otter instead of the itty-bitty Bird Dog. Eventually, Major Bartle became Commander of the 18th. I viewed our initial meeting as another one of those moments meant to be. He became influential in my future military service and an important mentor.

Nha Trang was a crown jewel assignment place. It was as far removed from war as you could be in Vietnam. The only other possible, more desirable location was Vung Tau, in the Delta Coastal region, which housed the only other Otter Company in Vietnam. Once, I had an occasion to fly there and spend the night. For total quality of life, Vung Tau held an edge. However, it was hard to believe that the lush officer living quarters in a downtown villa in Nha Trang were in a combat zone.

## Low, Slow and Reliable

The 18th Aviation motto was very fitting. Their radio call sign, *Reliable,* was known throughout the country. The 18th was the first fixed-wing Army unit deployed to Vietnam in February of 1962. Initially, it provided support throughout South Vietnam. When the 54th Aviation Company arrived, the Otter mission was divided between north and south.

The 18th headquarters and one flight platoon were located in Nha Trang. Two other platoons were located in Pleiku and DaNang. The company also continuously provided one Otter in Bangkok, Thailand. This aircraft supported the Joint U.S. Military Advisory Group. It flew missions similar to those in Vietnam to remote areas with short landing strips. The aircraft and crew in Bangkok rotated every 30 days.

My initial plan was to complete a transition checkout with the Otter at Nha Trang. Then, my subsequent assignment would be decided. In keeping with my past experience of unexpected events determining my fate, it happened again. A controversial inter-service agreement was signed between the Army's chief of staff and the Airforce. The future of Army Aviation was changed forever.

My first order of business was to get checked out in the Otter. I found it similar to the Beaver, with the exception of sitting higher off the ground and a higher nose sight picture with the final landing approach. The engine control quadrant was very similar. It felt very comfortable. After two days of intensive training, the instructor pilot signed me off as qualified to fly the Otter.

The operations officer said I would be flying an operational mission the next day as a pilot in command. The co-pilot was also a new guy, but he had not been checked out yet. This sounded to me like the blind leading the blind. The operations officer assured me the mission was a piece of cake. Years later, I remembered his statement when I saw the movie *Catch-22*. In this movie, the operations officer said something similar to a bomber flight crew before they took off on missions where half of them were shot down.

My first Vietnam mission is one I have never forgotten. It started out straight-forward. We were to fly to Bien Hoa, near Saigon, and meet a special forces major who would direct us to his desired destinations. We didn't even need a map! This relieved my significant concern for being able to navigate the remote locations in areas I had not seen. We found the major, a combat-hardened snake-eater kind of guy. There was nothing warm and fuzzy about him. In fact, he just seemed like a mean dude. He planned for us to resupply several A-team camp locations. We loaded up and took off, with him directing me where to go.

The first destination landing went smoothly. As we approached our next location, the runway looked odd. It appeared to be an old road leading to a large building. The landing strip seemed long enough and was paved. I was confident as I made the approach. However, I did not realize there was a significant dip in the runway. As we rounded out the landing, the Otter stalled and hit the ground very hard. Unlike the Bird Dog, the Otter did not bounce. It just slammed to the ground. We hit so hard that all the emergency parachutes fell off the overhead racks in the cabin area. My helmet hit the roof of the cockpit, and I questioned if I may have lost some tooth fillings. it was ugly. As we exited the aircraft, I was not anxious to confront the snake-eater.

He had already turned his attention to the cargo to be loaded and was engaged in discussion with the special forces soldiers who met the plane. I was relieved to see he had more important business than critiquing my landing. They loaded various boxes onto the plane, but I was unsure of what they contained. I had been told usual deliveries included ammunition, medical supplies, radio equipment, mail, and necessities for surviving in the wilderness.

Loading was just about finished when the major walked up to me. He drew his .45 caliber pistol, cocked the hammer, and pointed the pistol at my head. He was taking whiskey to the troops at the next camp. They were looking forward to it. He said if I landed at their site the way I did here and broke the bottles of whiskey, he would "blow my f_ing brains out."

He concluded with, "Do I make myself clear, Captain?"

I quickly responded, "Yes, Sir!"

As we flew toward the next destination, I thought about what the major said. I decided he couldn't possibly be serious about shooting me, but I was fully impressed with his ability to emphasize a point. The least I could do was make a smooth landing and avoid any further worries. He seemed to have a lot on his mind. I could tell he had a very tough job. Truthfully, I was scared shitless and knew not to make him angry again.

I set up a long, low, flat final approach like I perfected in Germany with the broken tailwheels. The touchdown was as smooth as grease. The cargo was safe. For the remainder of the mission, the major made no other comments other than directions. When we returned to Bien Hoa, I saluted him, which he returned and then walked away. I don't think I ever saw him again. Our flight back to Nha Trang was uneventful. I was relieved to complete my first combat mission. I told the operations officer that it was, just like he said, a piece of cake.

The headquarters for the 5th Special Forces Group was in Nha Trang, and responsible for all Green Beret operations in the country. It was commanded by a full colonel who traveled almost daily throughout South Vietnam. Much of his travel was by Otter.

During my short stay at Nha Trang, I flew missions for the colonel on two different memorable occasions. It is common for pilots to remember unusual flights. A popular saying during flight school said flying involved endless hours of boredom punctuated with occasional moments of stark terror. Personally, I found very few moments of military flying boring, with more than a few moments of terror! Overall, most of it was interesting, at a minimum.

## Goodbye Caribou

The inter-service agreement included the Army transferring its Caribou cargo aircraft to the Air Force as well as the Army foregoing any future use of fixed wing cargo aircraft. In exchange, the Air Force gave the Army full ownership of all future development and fielding of rotary-wing aircraft, including armed helicopters. The Air Force kept helicopters only for air-sea rescue missions. This agreement comprised a major clarification of growing roles and missional differences between the services. It became a crossroad in this journey that began 20 years earlier when the Air Force was established. The Caribous' transfer was to be completed no later than January 1, 1967. One Army Caribou unit was based at Qui Nhon, which would impact the 18th Aviation Company.

The potential long-term significance of this agreement has been lost upon most. Many Army personnel felt the Army was royally screwed with this decision. Later, I was told the Air Force made the agreement because they felt there was no future for armed helicopters on the battlefield. Today, the Apache helicopter is a primary means of close air support for our ground forces, while the Caribou is a mere footnote in military history. By the way, lots of Otters and Beavers still fly in Alaska and Canada!

This agreement resulted in a major power play in Qui Nhon over the facilities occupied by the Army Caribou Company. The 18th Company Headquarters and support maintenance detachment were moved from Nha Trang to Qui Nhon, in effect, to commandeer the facilities before the aircraft transfer was made. This would also keep the Air Force from using these facilities. I was selected to be the project officer to plan and carry out the movement of the equipment and involved personnel. The flight platoon would remain at Nha Trang, and everything else would go to Qui Nhon. The relocation allowed the large maintenance hangar at Qui Nhon to assist in the maintenance and repair of the Otters. This was also a more central location for the outlying platoons. Qui Nhon sat midway between the DaNang, Pleiku, and Nha Trang platoons.

During July and until the move was completed in August, I kept busy going back and forth coordinating the details. The Caribou Company administrative and living quarters at Nha Trang remained occupied during the interim. At Qui Nhon, we set up tents as temporary housing. Fortunately, the Caribou Company drawdown went more quickly than expected, which shortened our stay in the tents. The majority of our equipment was moved by sea and involved another set of players to coordinate with. Overall, it was a major undertaking that took all the time and energy I could give. The Caribou Company folks were not exactly happy campers during this period. It took a strong public relations effort to keep everyone working together.

After all the maneuvering to out-flank the Air Force, they ultimately had no interest in the Qui Nhon facilities and moved to a more comfortable air base to the south. They had different minimum acceptable runway length regulations for their cargo aircraft. The Army flew the Caribous into landing strips as short as 1,000 feet, as we did with the Otters. We flew to many of the same places, but the Caribou could deliver a much larger payload with a single sortie. With the change in the Air Force's minimum requirements, the Caribous could no longer use runways shorter than 2,000 feet. This caused a significant increase in cargo delivery for the Otters.

Speaking of happy campers, there was much talk and concern among the 18th troops about who would be the lucky ones to stay in Nha Trang with the 1st platoon. Although Qui Nhon was another coastal city, it was quite primitive compared to the French haven at Nha Trang. No one was anxious to move. Once the move was completed, I became the unit supply officer for the new headquarters at Qui Nhon. I saw this as an opportunity to continue to prove my worth to the Company Commander and hoped this would lead me to be considered for a platoon leader's assignment for the second half of my tour.

Major Bartle ran the company headquarters with great skill. He impressed upon each aviator the importance of developing our professional skills beyond flying. He sincerely showed interest in our full development as officers so we could keep up with our non-aviator counterparts in each branch. Major Bartle was a field artillery officer who also understood the challenges aviators faced to remain competitive within artillery. I learned a lot from him and have always been grateful for his leadership.

I remained in Qui Nhon until January 1967, when Major Bartle gave me the opportunity to command the 2nd flight platoon at Pleiku. Before my departure, we had a memorable holiday season. The company junior officers lived in a small civilian house the Army rented for us, just outside the airfield and very close to our workplace. The officers became friendly with our Vietnamese neighbors. We especially enjoyed the children. They learned our names and recited them when we came home in the evening. We had a fun Christmas party with them, which they enjoyed immensely. We also saw a Bob Hope Christmas show at the airfield alongside our hangar.

On New Year's Day, we staged an Army-Navy football bowl game on the soccer field in Qui Nhon. The Navy had a swift boat unit stationed at Qui Nhon. Since it was the coastal rainy season, we played in the mud and pounded the crap out of each other in a scoreless battle. On the last play of the game, the Navy team threw a Hail Mary touchdown pass and won the game 6-0. Three of us from the 18th were members of the Army team.

During my first six months in Vietnam, I flew a variety of missions from Nha Trang and Qui Nhon, which gave me the opportunity to see all of South Vietnam from the air. I flew the entire coastline from the demilitarized zone (DMZ) in the north to Can Tho in the south. It is a beautiful country. The beaches along the South China Sea are spectacular. I envisioned them as a prime future tourist destination, which is actually happening.

All of my references to Vietnam are actually about South Vietnam, as I did not have the opportunity to see any of North Vietnam. The only two aviators I know who flew over the north were both shot down and held prisoner in the Hanoi Hilton. One was an Air Force pilot, and the other was a Navy. Naval Officer Everett Alvarez was the first aviator taken prisoner when he was shot down in 1964. He remained a prisoner until 1973. I had the pleasure of serving with him on a special 2004 commission in Washington, D.C., to study the future of the Veteran's Administration healthcare system. The Air Force Officer, Don Heiliger, served as a member of the Department of Veterans Affairs (VA) Board when I was the Department Secretary in 2000.

In 1966, I saw two of my *Beaver* Mania buddies from Fort Rucker in Vietnam. Our singer, "Shine" Powell, flew Caribous. John Pfeiffer flew Mohawk surveillance missions over the Ho Chi Minh trail. John's unit was located at Hue-Phu Bai, and I flew a replacement engine to them from Qui Nhon in an Otter. "Shine" Powell was at An Khe with the 1st Cavalry Division, and I visited him there.

## Pleiku

Pleiku was a whole different world from Qui Nhon and a completely separate planet from Nha Trang. As much as the two cities were considered to be R&R centers, Pleiku was definitely a combat zone location. There wasn't much recreation at Pleiku. All the aviation elements based there were directly engaged in forward combat operations.

The main base camp for the Army's 1st Cavalry Division was at An Khe, situated between Qui Nhon and Pleiku. The main base camp for the 4th Infantry Division was slightly west of Pleiku at Camp Enari. The Military Assistance Command, Vietnam (MACV) 2 Corps Headquarters was also there as a Special Forces C Team. There were two airfields. The Pleiku Air Base was home to United States Air Force and Vietnam Air Force (VNAF) aircraft.

Army aircraft were parked at Camp Holloway. Overall, Pleiku was a very busy place. Pleiku's elevation is about 2,500 feet, with clear but cool winter months. This was a big change from the steamy coastal weather at the other bases.

In November 1965, the 1st Cavalry Division engaged a major North Vietnam Army (NVA) regimental force in the Ia Drang Valley, southwest of Pleiku. The battle was widely publicized in the book *We Were Soldiers Once and Young*. The book was turned into the movie *We Were Soldiers*. I feel it is the best account of an actual battle in Vietnam that I have seen. This film, along with *Lone Survivor* and *American Sniper,* is probably too factual for Hollywood's liking. Ridiculous portrayals like *Apocalypse Now* are more entertaining and appeal to those who embrace negative stereotype images of those who fought in Vietnam. The Ia Drang battle was as intense as any other battle in our history. The courage and heroism of the Americans who fought there deserve the highest respect.

*Stolen Valor* is a book that presents a well-researched account of how the myths and false characterizations of our Vietnam service warriors were created and sustained long after the war. The book explains how negative stereotype images of Vietnam veterans were manufactured by anti-Vietnam sources. It also explains how Vietnam vets became the most successful post-war generation. Unfortunately, I have been surprised and disappointed to learn how many Vietnam War phonies there are.

## The Otter Platoon

The Pleiku platoon had four assigned aircraft with a total of about 20 personnel, including pilots, crew chiefs and maintenance personnel. Daily, we were required to provide a minimum of two aircraft to the Military Assistance Command, Vietnam (MACV) and one to the Special Forces. In the Special Forces structure, the group headquarters is similar to a brigade; the C Team is similar to a battalion, the B Team is comparable to a company, and the A Team is much like a platoon.

We flew established north and south courier routes for MACV. Each day, the Special Forces headquarters provided support to one of their three B Team headquarters located at Kontum, Qui Nhon, and Ban Me Thuot. Each of those teams controlled a group of A-Team locations in remote areas. Many of the landing strips were unpaved and short, with the 1,000-foot runway at Plei Me as the smallest. This was a Special Forces camp located south of Pleiku, attacked shortly before the Ia Drang battle.

Shortly after I assumed command of the Platoon, the Camp Holloway Special Forces C Team Commander invited us to move into their compound. We had been living in the MACV compound, located at the main air base, but it was inconvenient for our aircraft and daily mission preparations at Camp Holloway. I accepted the offer. This was a good arrangement. Living with the Special Forces folks gave us a closer understanding of their culture and the bond of comradeship they shared. The relationship between officers and enlisted soldiers was much closer than conventional units. It helped me understand the concept of a Band of Brothers long before the phrase became popular.

The acronym SOG, or Studies Operations Group, was mentioned among them in hushed tones. For a long time, I was unaware of the classified missions they performed. Today, these missions are detailed in public reading. A series of books authored by Special Forces Veteran John Plaster provides great SOG information. A few years ago, I had the pleasure of taking him for a ride in my Wisconsin friend's Bird Dog. His stories and photographs from along the Ho Chi Minh trail in Laos are amazing.

Each day brought new adventures among our standard routes. Most of the coordination for each sortie was handled by Special Forces and MACV personnel in a smooth manner. But there were often surprise developments. We frequently received enemy fire during landing and takeoff. Occasionally, we were hit. We joked that the enemy saved their bullets for helicopters and spared us most of the time so they could catch a ride. There were usually people hanging around the airstrips hoping to become a passenger. We took them aboard if we had room and enough weight allowance remaining to get off the strip safely. I recall occasions of ejecting passengers before starting a takeoff roll after reconsidering the distance needed versus availability. I am fairly certain there were times when the Viet Cong (VC) got on board, but I was never highjacked.

It is hard to describe the unique and extremely remote conditions of many of our destinations. Ban Don was near the Cambodian border with a population of water buffalo and elephants in the area. One of my favorite snapshots is of these village children and their baby elephant. Montagnard tribal villages were even more unusual and impressive. Their distinctive structures were on stilts. I have seen nothing since that compares with the variety of third-world living conditions I witnessed in South Vietnam.

While I thought I had experienced the poorest flying weather in Germany, some of what we dealt with in Vietnam was worse. Continuous monsoon activity, torrential rains and limited ceilings and visibility made flying dicey. Their coastal monsoon season is during our winter months, and the inland monsoons are during our summer months. Fortunately, I arrived in Pleiku in January 1967 in clear, sunny, and cool weather. Actually, we wore flight jackets to stay warm. Many of the missions flown from Pleiku were to coastal locations with monsoon conditions. Every mission included challenges and risks. Our missions were single ships with a considerable distance from our base. Radio communications were spotty, and navigation instruments weren't much help. Many of the places we had to reach had neither voice nor navigational radio assistance. Our aviators did a lot of dead reckoning or "seat of the pants" navigation. During the poorest weather, we flew the coastline whenever possible because there were no terrain obstacles. Traversing inland included the constant threat of mountains. Many aircraft fell victim to these.

Dragon Mountain is near Pleiku. When I left in June 1967, there were eight different aircraft crash sites on it. Two months later, a Pleiku Platoon Otter struck Dragon Mountain, trying to return to Camp Holloway and killed the four persons aboard. The pilot in command was very experienced and someone I'd flown with numerous times. This tragic accident is a stark example of the dangers involved with this kind of flying.

When I returned to Vietnam in 1970 and served as an attack helicopter pilot, I frequently thought about the contrasts to my Otter pilot experiences. In some respects, the cross-country Otter missions were more dangerous than being in daily firefight situations, which covered a limited geographic area and an average flight time to the target area of only 15 minutes.

## The Home Front

Whenever I thought I'd had a bad day, I remembered Jackie and the boys in Tucson and their difficulties. They had no relatives or friends nearby other than Jackie's sister, Pat. Many military families experience bad things when the spouse is gone. Dave fell off a backyard fence and broke his arm. Mike fell onto a wicked Arizona cactus. A thorn embedded itself into his leg and had to be surgically removed. Other illnesses and crises of various kinds occurred throughout the year, as they also dealt with my absence and the stress of whether I would return home safely. There were no other Army families in the neighborhood but several Air Force families because of the Davis Monthan Air Base. The Air Force families had their own network and mutual support group. Many of their husbands were in Vietnam also. Jackie became friends with one spouse. It helped her to know someone else living a similar experience. I met her husband in Vietnam during a mission I flew into Bien Hua.

Jackie and Pat took the kids on lots of outdoor activities. One of their favorite activities was picnicking at nearby Mt. Lemon Park. They also went to Tombstone, where the kids reenacted the gunfight at the O.K. Corral. The most difficult thing we all faced was not being able to talk to each other. A year is a long time. Our regular letters to each other helped, but not being able to hear Jackie's voice was very hard. In 1967, cell phones were unheard of. With Military Auxiliary Radio System (MARS,) long distance calls could be placed through a ham radio phone patch system. It was extremely difficult to place a call this way. Part of the problem was the time difference. We discovered that on some nights around midnight in

Qui Nhon, we could connect to the Elmendorf Air Base in Alaska using the low-frequency single-sideband radio in the Otter sitting on the parking ramp. A couple of times, I got through to Jackie for a few brief moments. It was wonderful. The reception crackled. We had to say "over" after each question and response. It was a lot of "I love you, over" and "I love you too, over," but it was great.

## Where Will I Go From Here

This was a common question among all soldiers as the year in Vietnam passed by, and we awaited reassignment orders. For captains, the preferred option would be to attend our branch officer advanced course, also known as the Career Course. For me, this would mean returning to Fort Sill, Okla., for the 10-month Field Artillery Officer Advanced Course, considered a prerequisite to become eligible for promotion to major. My duty performance was praised by my superiors in the required written evaluations, as well as other commendations I received. My performance was definitely trending upward. I had worked very hard in every responsibility I was given. I felt that I was living up to the career counseling advice I had received in Germany. Things had gone so well. I had almost forgotten that I was still a National Guard Officer on loan to the active Army. The people I worked with did not see me this way at all.

Reality returned when we each received our orders. Other

captains in the unit were sent to their branch advanced course, but I was assigned to Fort Rucker as an instructor at the aviation school. I was disappointed but not surprised with this news. It was just another reminder that I needed to do more than a good job for me to succeed on active-duty. I didn't know what would get me beyond this. Major Bartle advised me that I needed a regular Army (RA) appointment to become fully competitive.

Now, I realize that when I returned to Fort Rucker, I faced *another* major decision. Would I leave active-duty voluntarily or remain and take my chances on the long-term possibility of being released involuntarily before reaching 17 years of active-duty? Remaining on active-duty would most likely mean a return tour to Vietnam as a helicopter pilot. I had no idea how difficult it would be to receive an RA appointment or the time frame this required. Clearly, I faced another crossroads.

## Homeward Bound

After 364 days and a wake-up, it was time to go home. As my final hours in Vietnam slipped away, I thought about how relieved and happy I would be to see my family again. I felt a sense of pride for having served my country during a time of war. I had no regrets or ill feelings about protestors or the anti-war sentiments in the U.S. I knew I was going back to the greatest country in the world. Everything I'd seen during the past year made me more grateful than ever to be an American. Those thoughts overshadowed any concerns I felt about the future or my career.

The flight back to the U.S. was very different than the one to Vietnam. As the aircraft lifted off the runway, a spontaneous cheer from everyone aboard brought most of us to tears. These were tears of joy that made each of us realize we were finally going home. When the aircraft touched down in California, the cheers were even louder. For most on board, their involvement in the war was over.

Few things can match the joy of seeing your family again after being apart for a year. I was amazed to see how much the boys had grown. Jackie looked more beautiful than ever. It was a thrill to be together. I could not imagine what it would be like ever to leave them again. We took a much-needed family vacation before I reported for duty at Fort Rucker. Leisurely, we traveled through Las Vegas and Utah. We were in awe of the spectacular scenery we found at the Zion and Bryce Canyon National Parks. Eventually, we made our way to Wisconsin and Chicago for family visits before we turned south towards Alabama.

As we approached Ozark, Ala., I could not help but think how it had been only three years since we had left there. I was overwhelmed with all that had happened in between. Once again, another unexpected bugle call sounded.

NORTH VIETNAM
DEMARCATION LINE
QUANG TRI
THUA THIEN
CAMP EAGLE
Ray - 1970
QUANG NAM
I CORPS
QUANG TIN
QUANG NGAI
KONTUM
BINH DINH
PLEIKU
Ray - 1967
PHU BON
PHU YEN
II CORPS
Ray - 1966
DARLAC
KHANH HOA
QUANG DUC
TUYEN DUC
NINH THUAN
PHUOC LONG
BINH LONG
LAM DONG
TAY NINH
III CORPS
LONG KHANH
BINH TUY
BINH THUAN
KIEN PHONG
CHAU DOC
KIEN GIANG
VUNG TAU
CAPITAL SPECIAL ZONE
KIEN HOA
VINH BINH
BA XUYEN
BAC LIEU
AN XUYEN
IV CORPS
THAILAND
LAOS
CAMBODIA
PHNOM PENH
GULF OF SIAM
SOUTH CHINA SEA
SOUTH VIETNAM
ADMINISTRATIVE DIVISIONS AND MILITARY REGIONS
JUNE 1967
International boundary
Province boundary
Military corps boundary
National capital
Province capital
Autonomous municipality

# Chapter 10

# Home From War

I never anticipated there would be another time in my life that would compare with the importance of the World War II era and the 1940s decade. I felt the way America changed and how I was affected personally could not be matched. Survival threats to the U.S. as a nation gave Americans a sense of purpose that was previously missing. A strong feeling of "united we stand" enveloped the country and drew Americans together.

During the 1960s, the opposite happened as this country became more divided than united. A historic combination of events took place which shook the very foundation of our nation's greatness. In my opinion, we have never recovered. Our country has been forever changed. In the decades after the 1960s, it is interesting how many different U.S. presidents have been accused of dividing America. There is plenty of blame to spread around. I think most of the damage was done in the 1960s with a complex combination of factors involved.

Among these contributing factors was the war in Vietnam. Personally, I reject the commonly held belief that the war was the defining event of this decade. If anything, the attention the war received may have served as a major distraction from the other important movements that took place during this time frame, which I feel really divided our country. I cannot see evidence, then or now, of any cause-and-effect relationships between the war and most of the events that occurred on American soil.

There should be little doubt that the most important event of the 1960s was the civil rights movement. My earlier experience living in Alabama exposed me to the disparity between blacks and whites. Dr. Martin Luther King, Jr. was viewed by many at the time as a divisive figure. Yet, a careful look at his messages draws the conclusion that his message was one of unity. Dr. King advocated for equal opportunity for all Americans, not just ethnic minorities. Unfortunately, his full vision for America was ignored and translated into subsequent political actions completely opposite of what he tried to promote. His goal of achieving economic equality and lifting people from poverty was replaced by a war on poverty that subsidized the problem rather than removed it. This created a permanent underclass sentenced to an endless circle of poverty.

President Kennedy's admonition to the country during his 1961 inaugural address begged Americans to "Ask not what your country can do for you. Ask what you can do for your country." This idea quickly evaporated when President Lyndon Johnson succeeded him. Johnson's concept of a Great Society completely reversed the Kennedy charge and launched an era of government entitlements that threatened to bankrupt the country even today. “Ask not” became “Demand now,” what your country should do for you.

Dr. King’s tragic assassination, along with those of President Kennedy and his brother Robert Kennedy, stunned the country and brought a state of chaos to our national morale and spirit. These untimely deaths were heartbreaking and weakened our faith in our strength as a nation.

The counterculture campus drug movement was another important dividing force in the 1960s. Radical campus thought and behavior were a phenomenon most Americans were not used to, and many could not understand. Much of the campus's anti-establishment energy was aimed at the Vietnam War. I feel there was little evidence of any substantive ideological basis for it. Once the military draft ended, anti-war protests ended abruptly as well. I have never understood why college students were exempted from the 1960s draft. This was not the precedence set in earlier wars.

While President Johnson is my personal choice for dividing the country, there are strong cases for others as well. There are those who believe President Obama divided the country, while many others believe his predecessor, President Bush, was the great divider. President Trump has been accused of the same. I believe the Johnson administration’s decision to exempt college students from the draft and not mobilize reserve component units into service increased reliance on drafting less advantaged young people. I see these historic disconnects as more divisive than the war itself. I never understood the college campus draft card burning protests by those least likely to be drafted. They may seem fitting in terms of the bizarre state of so much else at the time and the irrational state of campus extremism.

The growth of the feminist movement and equal rights for women also fueled the 1960s social revolution. This activity had nothing to do with Vietnam. As the 1960s ran its course, television news found a daily home with the Vietnam War and did its best to convince viewers every evening that it was the defining event of the time. The death of thousands of Americans was a national tragedy that deeply affected everyone. Whether one agrees with the way the war was reported or not, there is no argument for the terrible size and scope of our casualties. The dramatic nature of death on the battlefield diverted Americans from the less obvious fact that the death of a nation was a sickness taking shape in America. This illness continued long after the war.

## Back To College Again

Despite the turbulence of this crazy world, 1967 was a time I had to deal with important personal decisions. Jackie and I labored carefully over our future. We knew there were no easy answers to the questions we considered. An early conclusion we reached was our lack of a place to call "home." This can be a shocking reality for military families to discover. Our parental families were in two different locations. At this point, neither location appealed to us. I had been removed from teaching long enough that it no longer felt like something I wanted to do. My only other real qualification was flying. I'd considered this as a potential career since flight training school. Yet, I had one more year of obligated service. Another factor was my lack of a college degree. No matter what career direction I chose, completing an undergraduate degree was imperative.

The question of attending the Army Career Course was clarified when I was informed that I was deferred from attending the advanced officer course at Fort Sill for one year. The plan was for me to go there in 1968. My assignment to Fort Rucker was only for one year. Then, we would move again. Completing a civilian education seemed unrealistic. I estimated it would take at least a year and a half of full-time study to complete a bachelor's degree, depending on how many previous credits were transferred. If I remained on active-duty, it would be more than three years before I could attempt full-time study toward a degree. By then, it would probably be too late for this to benefit my future Army service. The rumor was circulating that when the Army started to draw down from the Vietnam War, any officer without a bachelor's degree would be released.

Another variable in the equation was whether I would be selected for promotion to major. I was in the zone of consideration for the next selection cycle. It looked like it could go either way. The Army faced a shortage of aviators. Maybe they would keep me around a while longer.

We settled into on-post family housing at Fort Rucker. I started my new job as a flight instructor. Then, I learned about a college extension program offered on-base through Troy State College, Ala. In the process of finding out what the college offered, I encountered another person who had an important impact on the rest of my life. Truly, it was another bugle call.

The full-time counselor for the Fort Rucker education office was amazing. This woman had a passion for helping every person as much as possible achieve their education goals. She worked harder to do this than anyone I could imagine. After a brief discussion in somewhat of a First Sergeant style, she TOLD me to start filling out forms. I was reluctant and said it seemed futile because I only had a year before I would leave Fort Rucker.

This person did not accept "no" for an answer. She said, "Boland, fill out the forms and quit wasting time. Someone in your situation can't afford to waste a minute. Get with it!!"

She needed my previous college transcripts literally the next day. By the time she finished reviewing everything, the counselor had a plan for me to complete all the bachelor's degree requirements from Troy State in exactly one year if I started the following week. I did not see how this was possible. She told me I didn't need to see anything; all I had to do was work my tail off for a year and get it done. "It's up to you, Boland," said this woman, who called everyone by their last name. "If you believe it's as important as you said it is, then prove it!"

For the next year, I put forth more work than I ever had up to this point, including combat. The plan was a mind-boggling combination of correspondence courses, written course competency exams, and resident courses. My daily schedule included teaching aviation tactics classes to flight school students during the day and attending college classes every Monday through Thursday from 5 p.m. to 9 p.m. The college classes were held in the same classroom buildings where I taught during the day. Every Friday evening, I performed required flight missions in support of student training. On weekends, I studied. Jackie and I realized that no matter what happened in the future, I urgently needed to complete this degree. We just had to do whatever it took for me to complete it. I had just returned from Vietnam. For another year, we had very little time for family enjoyment.

It was a grueling year. Several times, I stayed up all night studying for exams and worked a full day before taking the exams. It was stressful for our entire family and an experience I would not wish upon anyone. On the positive side, I had very good college instructors. Most of the classes were interesting and enjoyable. My degree program included a major in sociology with minors in psychology and history, all areas I personally found useful.

True to the saying that no good deed goes unpunished, I received a significant blow. During my final trimester of resident courses, Troy State completed an accreditation process to receive university status. All degree programs and course requirements were reviewed. I was informed by my steadfast counselor that my degree program now required three additional credit hours. I could only complete my degrees through resident study. I think she was actually more distraught than I was. It was a real punch in the gut for both of us, and I felt like vomiting.

About the same time that we received this news, Jackie's mother passed away suddenly from a heart attack. My mother-in-law was an angel who had treated me like her own son from the day Jackie and I were married. She had been there for us in the early days when we had nothing but each other, always encouraging and supportive. Our decision to be far away from family was painful for her, but she never complained or questioned our decisions. In the years ahead, Jackie, the boys and I missed her unconditional love terribly. Two weeks later, I was promoted to major. We knew Jackie's mother would have been very proud of how Jackie and I achieved this milestone together as a dedicated team. We felt a sense of accomplishment and realized there were other milestones we were capable of reaching.

## The Decision

As I kept up with the grind of my weekly schedule, I explored job opportunities and submitted several applications and resumes to potential employers. I had a positive reaction from a commercial airline that broke down when they realized I was not multi-engine rated. If I had been qualified in either the Army's Caribou or the Mohawk aircraft, it would have made a major difference. A fellow instructor with Mohawk experience was hired by a major airline and completed a full career with them. My lack of a college degree was a showstopper in every other opportunity. I received no offers. By now, I needed a bachelor's degree to teach. I would have also needed to attend school full-time. Without full-time employment, it would have been very difficult to attend school full-time and support my family. Health insurance was a major need, just as it is today.

Jackie and I weighed all possible scenarios. It became clear that the most sensible course would be for me to remain on active-duty, do everything possible to succeed as a career officer, and hope for the best. This choice was a practical one but with lots of question marks. My promotion to major was significant. Yet, we understood this came with no long-term guarantees. I still faced plenty of mountains to climb and obstacles to hurdle. As I went through the final stages of plotting out my future plan, I realized more and more that few things in life are guaranteed. Everyone runs the risk of losing their job on any given day. I accepted the idea to simply do my best and not worry about things outside of my control. This was the only way I could maintain my integrity and humility and still feel honest with myself.

As I look back, I see what an important shift this was in my thought process. I embraced the philosophy of engaging each day with an awareness of what I was prepared to lose rather than what I expected to gain. Armed with this attitude, I gradually became more relaxed, confident, and productive. Most importantly, I realized how much I liked being a soldier and serving our country. Every day, even during the controversial times of the Vietnam War, I was proud to wear the uniform. This special sense of pride and purpose really keeps a career soldier committed and willing to serve.

With this decision, the looming cloud became a guaranteed trip back to Vietnam in 1969. Meanwhile, we had another year to prepare for this. We embraced more family time and fun than ever before while I attended the Artillery Officers' Advanced Course.

## Fort Sill Round Two

Technically, this was my third stint at Fort Sill and second assignment as a family. Eventually, there would be one more. During this second round, we found on-post student family housing right away. This alleviated the usual major concern of finding a school for the boys. At the beginning of the 1968-69 school year, Dan was in fifth grade, Dave in third, and Mike at home. The older boys signed up for organized football. The previous year, they participated in Little League baseball. Life on an Army post offers a lot to families. Fort Sill was a self-contained city with schools, shopping, and recreation. The youth sports opportunities were extensive and enjoyable. Dave was always bigger than other kids his own age, so he played football with Dan's age group because of his size. Both boys showed athletic potential. At the time, I did not foresee they would complete high school during a later assignment at Fort Sill. As seniors, they were both the starting tight end for the Lawton Eisenhower High School football team.

Volunteers coordinated most on-post youth sports. Over the years, I coached several basketball teams. The Fort Sill area is an interesting and historic area; there is much to see and do. As a family, we made the most of it. The nearby Wichita Mountains Wildlife Refuge had a herd of buffalo, making it a unique place to visit and photograph. The area is rooted in the American Indian history of the Comanche and Apache tribes. Fort Sill is the burial site of the Apache leader Geronimo.

Most of the Artillery Officers' Advanced Course students were captains and regular Army officers. Many were also West Point graduates. I was one of the few National Guard officers. Because I was a major, I was designated as a section leader. Within our class, students developed close and long-lasting friendships. Two of my closest friends became General officers. Vietnam was foremost in everyone's mind, and we knew that some of us would return for a second tour of duty. There were six Vietnamese officers in my class, and one of them was Major Tuan, who finished as the top student in the class. He was a brilliant guy, yet very humble and gracious. Jackie and I hosted the six officers at our home one evening for dinner, which included a memorable discussion about South Vietnam's future.

Major Tuan shared that he realized the Americans would eventually have to leave Vietnam. He predicted this would be followed by a major invasion by North Vietnam, which would have the full backing of China and the Soviet Union. South Vietnam would be defeated, and communist rule would take over the entire country. He said that South Vietnam could not prevent this from happening. Yet, he emphasized that the spirit of freedom and self-determination the Vietnamese people learned from the U.S. would not be taken from them.

The events actually happened the way he described them. He also predicted that sometime in the future, freedom would return to South Vietnam. Major Tuan expected this to take at least 50 years. He shared these predictions in 1968.

My current wife and I toured Vietnam in late November 2018. The group consisted of all Vietnam veterans and were mostly aviators who served in the Pleiku area. Christmas season had arrived. We were surprised to see Christmas trees in all the hotels and business places. We heard American Christmas songs everywhere, recordings by the original American singers. The Vietnamese people acted very friendly toward us. Many, especially younger people, spoke English. Some wanted to have their picture taken with us.

It was surprising to see such evidence of continued American influence. I found it startling to realize the positive long-term effect evident in the country from when Americans had been present some 50 years earlier. Everywhere I looked, I found Major Tuan's prophecy alive and well. The strongest testimony came from a young man I talked to in the lobby of our Saigon hotel where he worked. He said his grandmother had moved to the U.S. and lived in Houston. When I asked him if he would like to live in the U.S., without hesitation, he said, "Everyone would like to live in America."

## Transition Training

Knowing I would go back to Vietnam after the Officers' course, I asked the Washington D.C. assignment office to schedule me for a helicopter transition training enroute. Eventually, I needed to compete for advancement with contemporaries who had combat helicopter experience. The helicopter was the wave of the future for Army aviation. My request was granted, and I received orders to attend a 12-week rotary wing transition course at Fort Rucker. This course included qualification in the UH-1H or Huey helicopter. This qualification positioned me to seek command of one of the many Huey units in Vietnam. In 1969, majors commanded most Army aviation companies. This was a credential I needed to acquire to keep up with the pack.

Unfortunately, a few weeks before the artillery officer course graduation, I received very disappointing news. My orders were amended, which canceled my attendance at the UH-1H course. Instead, I was assigned to attend a National Guard OH-13 rotary wing qualification course with a similar reporting date. I promptly called the assignment office to see if there was a mistake. There was no mistake. My UH-1H quota was given to someone with higher priority. I argued there were no longer any OH-13 Sioux aircraft remaining in Vietnam. Qualification in this would be meaningless. The assignment office corrected my assertion and informed me that

I would still be rotary wing qualified. A UH-1H qualification was something I could acquire in Vietnam.

This development was another reminder of my second-class citizen status in the National Guard and how much harder I must work to change this. This was especially hard for me to swallow. While I had volunteered to become a helicopter pilot, many of my fixed-wing-only colleagues were still hiding in the bushes to avoid it.

I proceeded to Fort Rucker as ordered but decided the time had come to take matters into my own hands. I could no longer assume that if I did a decent job, the system would take care of me. In my case, it felt more like a joke. I felt little assurance the Army would have a long-term spot for me. I needed to come up with a plan and, more importantly, a way to make it work. I needed some creative thoughts.

During my first Vietnam tour, I met members of the 2/20 Aerial Rocket Artillery Battalion assigned to the First Cavalry Division or the Cav. They were friends of my former commander, Major Bartle and visited us in Qui Nhon. Lt. Col. Bartholemew was part of this group. He was a battalion commander who was later killed in action. The mission and success of this unique organization were well-known. The only other similar unit was in the 101st Airmobile Division.

I felt an assignment to either of these units was the perfect place for an artillery aviator to be. By the time I arrived at Fort Rucker for the transition training, I had a plan. My goal was to become a commander for an aerial rocket artillery battery unit in the 101st Division. I chose the 101st Division because I knew the Division Artillery commander was an aviator who was also well-known by the senior leaders at Fort Rucker who knew me. I hoped to persuade one of them to send a letter of recommendation on my behalf. The recommendation was sent, but I was cautioned there was no guarantee it would help my cause because my plan was such a long shot.

I completed the eight weeks of training, settled my family in nearby Enterprise, Ala., and prepared to depart for Vietnam. I still had no idea what my assignment would be when I arrived in Vietnam. I felt like I was flying blind into a fog, but somewhere, there would be a safe place to land.

# Chapter 11

# The Vietnam Screaming Eagles

Getting on an airplane for a return trip to Vietnam felt much different than my first assignment. This time around, I knew what to expect. I realized my assignment as a helicopter pilot would be more dangerous than the airplane missions that I flew in 1966-67. I also knew what it felt like to be separated from my family for a year. There was a nagging, empty feeling inside of me asking, "Why are you doing this?" I had no satisfactory answer, other than I felt an intense sense of duty.

Later, I tried to imagine the various emotions felt by soldiers who served multiple deployments to the Middle East. It's hard to describe. Each deployment never gets easier. Saying goodbye to my family was harder with the second deployment. Jackie and I decided that she and the boys would not go with me to the airport. We shared our final hugs and "I love you" at home, after which my friend John took me to the airport. We never spoke of the worries about returning safely in one year. I knew how difficult these thoughts were for me to bear. I can only imagine the daily stress Jackie felt.

By 1969, the U.S. military knew that we would not win the war. At some point, the Americans would leave, just as my

Vietnamese friend had forecasted. President Nixon announced a plan for Vietnamization, which meant a phased handoff of all responsibility to the South Vietnamese government and its armed forces. This process began after my arrival in November 1969, and continued until I left a year later. By then, all the 101st Division ground troops had pulled back and the 1st Army of the Republic of Vietnam (ARVN) Division was in full charge of the area. This division was considered one of the absolute best in South Vietnam. It did an excellent job in everything I saw and supported. Despite the withdrawal of our ground troops, ARVN continued to receive full helicopter support from the 101st.

My travel to Vietnam was uneventful. Upon arrival, I waited for my assignment at the replacement center. The usual bulletin boards still posted assignments. After a couple of days, I saw my name directing me to the 101st Division, known as the Screaming Eagles. With step two of my grand plan accomplished, my thoughts turned immediately to what would happen once I joined the Division.

Upon arrival, I was told to report in-person to Division Artillery Commander Colonel Moore. At the time, there were two men well-known as Colonel Moore in Vietnam. Colonel Hal Moore was the battalion commander in the 1st Cavalry Division, whose soldiers fought in the famous 1965 Ia Drang Valley Battle. I reported to Colonel Howard Moore.

Colonel Moore immediately put me at ease and welcomed me to the Division. He had heard some good things about me and laid out his thoughts about my circumstances. Initially, I would be assigned to the 101st Aviation Group Headquarters where I would have the opportunity to receive training and qualification in the UH-1H Huey aircraft. The 101st Aviation Group Commander, Colonel Sinclair, was aware of my situation. Moore advised me to do the best job I could in whatever capacity Sinclair wanted me to serve. As for the aerial rocket artillery (ARA) battalion, there would be command openings down the road. This unit was equipped with the AH-1G Cobra attack aircraft which required another 25 hours of training to become qualified. Moore was honest in his assessment. My chances of doing all this were slim, but possible.

After reporting for duty, I quickly discovered that Sinclair, like Moore, was an outstanding leader, mentor, and person. Under his tutelage, I began to realize the 101st Division was not just a place for me to fulfill my selfish goals; it was a place where I was exposed to a breed of Army culture I had not experienced. While I had served with many good officers, this leadership and culture was different. Everyone I encountered displayed top-notch standards and professionalism. Like many other aviators, I found the overall Screaming Eagle spirit contagious. Everything I'd read about the Division record from their history during World War II made more sense. I found myself actively living another chapter of its storied history.

Colonel Sinclair put me to work in his headquarters within the tactical operations center (TOC) as an assistant to the Group Operations Officer Colonel Ray Pollard, who was another superb officer. It almost felt too good to be true when I discovered that Pollard was slated to be the next commander of the ARA Battalion. All at once, my table was set. If I could prove to Sinclair and Pollard that I was worthy of advancement and continued development, my goal might be attainable. I was ready to work as hard as possible. Yet, I had no idea what challenges I would face to get there.

The tactical operations center (TOC) was responsible for the daily planning and coordination of missions for two assault helicopter battalions and one assault support battalion. A general support company provided daily aircraft for the Division headquarters staff, including the commanding general and two assistant division commanders. Each company-sized unit had a unique call sign. To this day, I still remember them all.

The TOC was located in an underground bunker. The activity pace was intense. The surrounding area received occasional enemy rocket attacks. Shortly before I arrived, the nearby mess hall took a direct hit.

I was blessed to work side-by-side with Captain Ken Chien, who tutored me on every aspect of our duties. Ken became a dear friend. Later, I had the pleasure of working with him on two subsequent occasions. Truly an amazing person, Chien was born in mainland China and taught Mandarin Chinese at West Point. He was assigned as a translator for the U.S. Embassy staff in China.

Evenings were our busiest time of the day as we worked to pair support with mission requests for the following day. Combat assault missions were flown daily throughout the area of operations (AO) to insert, reposition, and extract infantry units. The AO stretched from Da Nang northward to the Demilitarized Zone (DMZ), and westward to the Laotian border. This was known as Northern I Corps and included well-known battle sites like Khe Sanh and Hamburger Hill. A typical aviation support package was a 1-10-2. One command and control UH-1H Huey, ten UH-1H assault aircraft, and two AH-1G Cobra gunships. Approximately 50 percent of our assigned aircraft were in the air every day. It was a daily grind to get every person, aircraft and mission coordinated and fulfilled.

About two months into my tour, Sinclair assigned me to a new mission. He selected me as his Assistant Division Aviation Officer (ADAO) where I worked in the division operations center. This center had oversight responsibility for all division aircraft coordination. To acquaint me with the broader scope of division staff activity, Sinclair asked me to accompany him to an evening briefing held by the commanding general. This was an extremely eye-opening experience. The briefings began at exactly 5 p.m. when the first briefing officer announced, “Sir, the time is 17:00 hours.” Each briefing officer wore freshly pressed jungle fatigue uniforms and spit-shined boots. They delivered carefully prepared reports with extremely specific details of combat operations in the preceding 24-hours throughout the division zone. Like me, the briefing officers were majors. I was severely impressed with their overall professionalism.

Sinclair also clued me in about a breakdown in communications, a major issue he wanted me to resolve. There was daily confusion regarding the number of available aircraft for mission support, which led to a credibility problem for our headquarters.

In time, we were able to clean up the discrepancies in daily aircraft status reporting. Sinclair had me visit a number of different units so I could see first-hand the significant variances in status. Boy, was this an eye-opening education in aviation maintenance management procedures. Brig. Gen. Jim Smith, Assistant Division Commander for Support, was another remarkable leader. He personally taught evening classes to company commanders with specific procedures for how to properly supervise maintenance and spare parts supply management. Smith became one of my most important mentors. His influence continued well beyond Vietnam.

In a few short months, I gained complete perspective of

division-wide combat operations and an in-depth understanding of all aspects relating to helicopter employment and maintenance. With my lack of experience, I felt truly fortunate to accomplish this level of education and leadership training. The decision window about my potential assignment to the ARA battalion drew near. Soon, Pollard would assume command of the ARA battalion. I had completed the UH-IH Huey qualification training and was eligible to undergo Cobra training. Already, I looked forward to continuing to serve under Pollard. I found him an outstanding officer as well as a gentleman. Pollard informed me that he wanted me to assume command of the B Battery in early April. In light of this, he scheduled me to attend a two-week AH-1G Cobra qualification training course. There was just enough time for me to complete the training and return before the departing commander left Vietnam. Everything was set except for one small detail.

On the day before I was to depart for Cobra school, Sinclair summoned me. He had not approved the ARA assignment because he had other plans for me. He programmed me to assume command of a UH-1H assault helicopter company. Briefly, I felt flattered. I was the least qualified guy who was wanted for two different helicopter command jobs.

I expressed my most sincere appreciation to Sinclair but stressed the importance of accepting the command of a field artillery unit. Then, I held my breath. Sinclair paused for a moment, grinned, and said, "You artillery guys really stick together, don't you?"

I nodded.

He said, "Okay, go ahead and go."

I wondered if he had been yanking my chain, but I didn't ask any questions. I saluted the colonel and thanked him again for his reconsideration. He wished me good luck. Deep down, I knew Sinclair had ample respect for the ARA battalion. Once, he and I watched two ARA Cobras in close formation come in for a landing. He was impressed and remarked, "THAT's the way to make a good landing approach!" I felt embarrassed to skip out on Colonel Sinclair, who had given me so much opportunity and responsibility.

When I arrived at Cobra school, a warrant officer instructor pilot checked me in. He asked to see my flight records folder. After reviewing it, he looked at me and said, "You've got to be shitting me." I didn't understand his expletive remark. According to him, because I had such little helicopter training and zero helicopter experience, it would be impossible for me to become qualified in the Cobra in two weeks. The Cobra's advanced technology and weapons systems made it a completely different challenge to fly. Clearly, he was unimpressed with the 1,000-plus fixed-wing flying hours I had, including one year in Vietnam. He was also unaware that I'd flown the Cobra on a combat mission with the commander I was replacing and had operated the nose turret weapons.

In rebuttal, I shared that in three weeks, I was assuming command of an aerial rocket artillery unit in the 101st Division and needed to be Cobra-qualified. I intended to accomplish this and would manage the challenging training. Realizing I wasn't going to throw in the sponge, he informed me that I would start the following morning.

In the one mission I had previously flown the Cobra, I learned enough to see how different it was to fly. The flight controls are significantly different. The first thing my predecessor had me do was bring the aircraft to hover and hold it steady. Immediately, I discovered the amazing effect of the stability control augmentation system. It has like two-times the effect of power steering. When the system's components are engaged for pitch, roll, and yaw (known as "three in the green"), the very slightest control pressure causes aircraft movement. Once a pilot can effectively control the aircraft while hovering, flying it becomes much easier. These knowledge tidbits helped me a lot when I started Cobra school.

The Cobra's cockpit felt comfortable and reminded me somewhat of the L-19 Bird Dog. The tandem seating and narrow space were remarkably similar. I quickly adapted to flying it. The two weeks flew by quickly. I received a "well done" on graduation day. As I thanked the instructors for their help, I graciously chose not to say, "I told you so." I was ready to perfect my Cobra flying skills.

## El Toro

When I returned to Camp Eagle, Col. Pollard directed me to assume command of B Battery immediately. My predecessor, Major Fred Stubbs, had departed a week early. This was unfortunate because I had counted on having some overlap with him. He had an outstanding reputation, and I was eager to learn as much from him as possible. My one orientation flight with him before I went to Cobra school convinced me his reputation was justified. Pollard also shared the unfortunate news that one of the Toro aircraft had crashed and both pilots were killed. Everyone in the unit felt this loss deeply. I signed the assumption of command order on March 29, 1970, and went to work. The B Battery name was the "Toros." My call sign and identity for the next eight months was "El Toro."

Three days after I assumed command, on April 1, one of the Toro aircraft was shot down in the infamous Ashau Valley. The Hamburger Hill battle was fought there one year earlier, and during the four months I had already served in the 101st Division, I learned how deadly this area was. I knew immediately the downed crew was in grave danger.

When this happened, I was flying in the front seat of a Cobra piloted by the Battery Executive Officer, Captain Jim McGrory. It was my first flight since assuming command and Jim was giving me an orientation on our mission. We heard an emergency radio call from Toro 92 Charlie, who Cpt. McGrory informed me was WO Fred Cappo. Cappos' voice was rushed and emergent as he quickly reported he was descending toward the downed aircraft to attempt rescuing the crew. This brief transmission was all we received as radio contact was lost due to the mountainous terrain. We headed towards their location, and it seemed like an eternity before we heard Cappos' voice again. He said he was attempting to depart the valley with the downed crew hanging on the outside of his aircraft. He added he would attempt to reach Firebase Bastogne, the nearest location with friendly ground elements. We proceeded toward the same location.

Both aircraft landed at Bastogne at the same time, and I was forever grateful to see that all four of the pilots were okay. The downed aircraft had been flown by Captain Dick Femrite, and his co-pilot/gunner was WO Joe Maxsom. They had received a mission from our operations center to attack enemy trucks that were sighted by a fixed wing reconnaissance aircraft. During the attack, they received heavy ground fire and the lead aircraft flown by Femrite was struck. With oil pressure dropping rapidly, he attempted to head for safety but was forced to land. He stated their aircraft still had ammunition on board as well as its classified radio equipment. We called for a medevac aircraft to take both crews back for evaluation. Femrite volunteered to guide us back to the site when I was directed by my headquarters to assess the possibility of extracting the downed aircraft. He got in the front seat of another Cobra that had joined us, and our flight of two Cobras headed back to the Ashau Valley.

Once there, Femrite described the location and intensity of enemy groundfire they received, while we orbited overhead the valley at a safe altitude. It didn't take long for me to conclude an extraction was not feasible. I radioed back to Col. Pollard this was the case. Shortly thereafter, I was directed to attempt to destroy the aircraft with rocket fire from our two Cobras. I acknowledged the order, and we began to fire. On climb out from the first firing run, we began to receive groundfire and I could tell my aircraft was hit. The downed aircraft was hit but not destroyed, and we decided to try again. On the second pass, we were still unsuccessful, and the other Cobra was hit during climb out.

I radioed back to Col. Pollard again for an update. I told him that if we continued our mission, there was a strong likelihood that we would be shot down as well. I recommended we be permitted to return to base and that they request the Air Force bomb the area to ensure destruction of the Cobra and its contents. My request was approved, and we were able to make it back to Camp Eagle.

After only three days of serving as El Toro, it was clear this was not going to be a business-as-usual assignment. This was an intense war zone where anything could happen and often did. I thought of the saying, "Be careful what you wish for." Although this was an assignment I had dreamed of, it already looked like it could become a nightmare. Little did I know, the worst was yet to come.

## The Mission

The ARA Battalion had two other firing batteries: A (Alpha), known as the Dragons; and C (Charlie), known as the Griffins. Each battery was assigned 12 Cobra aircraft. There were three platoons in each battery. Headquarters Battery had three UH-1H Huey support aircraft. Each Cobra was configured with four large rocket pods, each having a capacity to carry nineteen rockets. Fully loaded, each aircraft could carry seventy-six rockets.

A variety of warheads were available to us, including high explosives, tear gas, and other warheads. One significant limitation was that the nineteen rocket tubes fired in a set sequence. Different rounds could not be selected from the cockpit when the Cobra was in flight. Warheads were loaded in order of anticipated need during each mission. The crew had to memorize the load sequence and manually keep track of each round fired so they knew which warhead types remained. Keeping track of all of this was daunting. The aircraft nose turret contained a mini-gun, and a 40-millimeter grenade launcher. It was operated by the front seat co-pilot/gunner.

The operating limitations of the aircraft were another crucial factor. Extreme heat had a negative effect on engine performance which meant we had to limit the aircraft's total weight for takeoff. To accomplish this, we reduced the typical rocket load from 76 to 60, reduced fuel load by 40 percent, and reduced the turret weapons ammunition load by 50 percent. Even with this configuration, we could barely takeoff and needed to refuel within an hour and a half after takeoff.

The B Battery's mission was to reinforce the fires of the direct support howitzer battalion for the Division's Second Brigade. Our secondary mission was to answer calls for fire throughout the division zone of operations. We also provided support for special forces missions conducted by the Studies Operations Group (SOG) headquarters at Combat Control North. Highly classified at the time, these missions involved the insertion and extraction of teams for reconnaissance of the Ho Chi Minh Trail in Laos. Extractions were often performed under emergency circumstances known as prairie fires.

Our mission was a 24-hours-a-day, 7-days-a-week responsibility. We were required to have a two aircraft fireteam in the air in two minutes, called a hot section. Crews stayed in a ready shack immediately next to the aircraft. We also had five-minute and 15-minute sections that were ready to move up when the hot section took off. Our battery had the constant requirement to have six aircraft and crews mission ready at all times and at least two others on standby. This arrangement was consistent with the field artillery pledge that its weapons are never in reserve and always ready to fire.

Our ARA Cobra's employment and control was unique compared to all other Cobra employments in the Division. The ARA Cobras had some pre-planned missions, but the majority of our missions were considered on-call. A typical call included responding to a call for fire from an artillery forward observer (FO) who was imbedded with an infantry unit that was in contact with the enemy. Situations were typically urgent. When our operations center received the call, the hot section was launched. Once in the air, aviators were given grid coordinate location, flight time to the target, and the radio contact information for the FO. The total duration, from when the initial call was received until the aircraft arrived on station, averaged 10-15 minutes.

The FO would use some method to identify the exact target location. The lead aircraft would fire two marking rockets. Then, the FO provided adjustment instructions in meters (left, right, add, subtract) and when to fire for effect. The lead aircraft commander determined the number of rockets to fire and began the attack. A steep dive was normally used to keep the burst radius on the ground as tight as possible to prevent injury to the friendlies. The second aircraft would attempt to have rockets hitting the target as the lead aircraft was climbing out. Each aircraft would fire its nose turret weapons to provide suppressive fire during their climb out. Prior to making the second run, the FO stated any additional adjustments or repeat fire for effect.

This process continued until the FO announced, “End of mission,” or some other option. Sometimes, the size and danger of the enemy attack were so great the FO requested the Cobras continue firing until all our ammunition was expended. When this situation appeared likely, we radioed back and had our next section launched and continued the fight while we rearmed and refueled. On some occasions, this rotation process lasted for several hours.

## Mission Training

To become proficient in everything I’ve described, each pilot completed an extensive training program. This involved a progression, starting in the front seat of the wing aircraft and culminating to the back seat of the lead craft, designated as a flight lead. The total time to complete this training varied but three months or more was common. Aviators had to learn much more than just flying and shooting. Just radio communications were a handful for a flight lead, who frequently listened to multiple radios simultaneously and switched among them to transmit. Aviators used intercom communication with their copilot, air-to-air communication with the wing ship and other Army aircraft involved in the mission, US Air Force aircraft, the FO, the base station, etc. It was often very hectic and demanding.

As the unit commander, I felt the need to accelerate my training process and become effective as quickly as possible. This made the unit warrant officer uncomfortable. No none particularly wanted to fly with the commander. I can't say I blame them. I respected their skill and proficiency very much. To be an effective commander, I had to be on missions, learning first-hand what my troops were up against and how I could best support them. There is no question a significant amount of stress accumulated amongst us. Just trying to get through each day was terribly draining. Mostly, we did not realize how the daily stress kept accumulating and building up.

Although we withstood a lot of hostile fire during our missions, the far greater threat was indirect enemy fire hitting our base camp. Mortar and rocket attacks were frequent. Camp Eagle was a sprawling complex located in low-lying terrain near Hue and Phu Bai. Our heliport was located in the camp's center and was known as the Bull Pen. We were the primary reaction force for base defense. Nearby enemy forces knew us well. Intelligence reports confirmed we were the primary target in the camp.

At approximately 3 a.m. on May 3, 1970, we underwent a heavy rocket attack that caused major damage and destruction to our aircraft. A barrage of seventeen 122 mm rockets hit the Bull Pen. One was nearly a direct hit on one of the hot section aircraft. The blast rotated the Cobra so it pointed directly toward our unit area.

As it burned, the fire caused rockets to cook off all the ammunition in the hot section, including machine gun bullets, and fire directly toward our unit area and beyond. It was a spectacular fireworks display. One rocket hit the maintenance building, set it afire, and destroyed one aircraft inside. The helipad sat slightly higher than the rest of the area. Most oncoming rockets flew over the top of the buildings. Many went directly over my headquarters/sleeping quarters or hootch.

When the rocket fire began, I rolled off my bunk, onto the floor and grabbed my flak vest. The rockets broke the sound barrier as they passed over my roof. I had no idea what these continuous blasts were. I became totally confused and tried to exit the hootch's front so I could get to the outside bunker. Another blast stopped me. I turned around and crawled in the other direction. Suddenly, the blasts stopped but bullet fire continued. It sounded like groundfire. I froze and feared a massive ground attack on the base was underway. I was terrified we would be overrun. Fortunately, there was no such attack. We survived with only one seriously wounded soldier. I'm confident everyone who was on Camp Eagle that night has never forgotten this attack.

When all the excitement quieted, we realized our beloved Bull Pen was in shambles. Almost all the Cobras were destroyed or severely damaged. Our maintenance hangar was burned to the ground and our operations center was destroyed by our own rockets.

It was a shocking and devastating sight.

Fortunately, five of our aircraft were off-site that night. When these aircraft returned, they were unable to land due to the debris covering the Bull Pen. Instead, they landed at a neighboring unit.

The next night, May 4, 1970, tragedy struck again. Two of our remaining aircraft were on two-minute status when they were launched to support troops under attack at Firebase Katherine. A UH-1H flare ship illuminated the target area while our Cobras flew below them. A flare got hung up on the UH-1H during release and exploded. As the burning wreckage of the UH-1H fell, it struck one of our Cobras, causing it to crash and burn as well. While this tragedy seems impossible, it happened. Soldiers on the firebase witnessed this awful event. Everyone aboard both aircraft was killed in action.

The terrible impact of these two nights was a severe blow to morale and confidence among the Toros. My leadership abilities were stretched to the limit. It took true grit to keep going and leading. There was an understandable sentiment among our unit, as well as our chain of command, to stand the unit down completely and distribute our personnel to other units. After careful consideration, we decided to stick together and continue on. I believe the division felt it was necessary to retain the Toro's outstanding combat record and cohesion. It was decided to do everything possible to re-equip and rebuild the unit as quickly as new aircraft and maintenance equipment could be acquired.

Our unit was re-constituted more quickly than I expected. We remained an effective fighting force throughout the rebuilding process. As the days and weeks passed, all officers and enlisted people in the unit pitched in to get our base repaired. It was a massive effort. To this day, I'm in awe of what they accomplished. The setbacks we suffered gave way to a new sense of purpose with the desire to not only succeed, but to excel. And excel, we did! During the recovery effort, the unit sustained the best record of total monthly flying hours and aircraft availability of any Cobra unit in the Division.

In the following months, there were more rocket attacks. Their constant threat made it hard for everyone to sleep at night. I saw the growing effects of sleep deprivation among our pilots, many of whom were on nearly constant mission alert status. The largest contributor to individual stress was the lack of regular and consistent sleep. On one occasion, I recall ordering a pilot to stand down for a couple of days after flying daily missions for twenty-one straight days. He tried to resist my concerns by saying he was still good to go. Warrant Officer Malcolm Hackney was only 20 years old.

I also felt myself wearing down. More than once, I fell asleep and slept for nearly 24 hours. Other times, I lay on my bunk, staring upward and wondering when I would take a direct hit. This came close to happening when an enemy rocket landed next to my hootch and blew apart the back end of it, which was another exciting night.

In July 1970, the 101st Division had its last major battle at Firebase Ripcord. At times, every available ARA aircraft provided fire support during the battle. Pollard and I joined efforts and flew two of the aircraft. Our Ripcord troops were pounded with relentless rocket and mortar fire until they were finally extracted. As bad as we thought we had it at Camp Eagle, it was nothing compared to Ripcord.

When I reached the end of my time as El Toro in November 1970, only two of the original 12 Cobras we had when I started were still there. It had been a long eight months. I was very tired. I lost fifty pounds during this year in Vietnam. As I departed the Bull Pen on my final day, I asked my Jeep driver to pause for one last look. At that moment, I thanked God for allowing me to survive and for giving me the honor of commanding the most brave and dedicated officers and men I have ever known. I also promised myself that I would never again take any of God's blessings for granted. Every day, for the rest of my life, I would try to honor and love my wonderful wife and sons by striving to be the best husband and father possible.

I knew the year had changed my life in multiple ways. At the time, I didn't understand exactly how much. A few days after I returned home, my wife told me that I had changed. I needed time to recover. I denied having any problems and tried to convince her, and myself, that I was fine. None of us coming out of Vietnam had any clue what post-traumatic stress disorder (PTSD) was. We simply were glad the war was over for us, and we were home. Memories of serving as a screaming eagle followed me for a very long time.

# Chapter 12

# The Cav

Jackie was right. It took time for me to wind down from the stresses of combat. I was jumpy and irritable. I had no idea how wired I was from what I'd been through, and another year of family separation had taken its toll. I questioned how much my family really needed me. Dan was 12 years old. He and his brothers had grown remarkably close to their mother. They were her big-boy household helpers. Jackie learned to deal with things on her own and did a wonderful job.

I felt left out. Being absent for two of the last four years, I didn't realize what my family had been through without me. In light of my experience, it is difficult for me to picture what returning prisoners of war (POW) experienced after they have been absent for much longer periods of time. When a person is gone for an extended time, it's easy to understand why many marriages collapse. Another example of the many hidden outcomes of war, these challenges receive little attention among the general public.

I returned to duty at Fort Rucker in November 1970 and was once again, assigned to teach airmobile tactics to flight school students. My personal priority was to complete the three remaining course credits required to receive a Bachelor of Science degree. In 1971, I finally received my diploma. This was only 16 years after graduating from high school, and I breathed a deep sigh of relief. For the next several years, there was a major reduction in military numbers. Many of my National Guard comrades were involuntarily released. Unless they returned to National Guard service and were able to complete 20 years of total service, they were not eligible for retirement benefits.

Gradually, we began to resume a more normal family life. My schedule was much more manageable, with five-day work weeks and no night school. This provided lots of time for family activities. The renewal of trips to the beach was a favorite with the boys. Slowly, I felt myself unwinding. As this happened, I started to feel better about life. Being able to relax was great therapy and I could tell it was working. I had survived ten exceedingly long years of active-duty service; a period that almost felt like a lifetime. Now, it was time to refocus on the future and contemplate my next steps.

To remain competitive for advancement and another ten years of active-duty service, I needed to be selected to attend the Army Command and General Staff College. Completing these studies would keep me in the running for promotion to lieutenant colonel. Without a regular Army appointment, these steps were nearly impossible. I submitted a new application, and it was disapproved. When this happened, I knew it would take a miracle for me to remain on active-duty for 17 years, which would lock me in for 20 years of service and retirement benefits.

One year later in April 1972, I heard another unexpected bugle call. Its announcement signaled a possible miracle. I didn't realize the whole meaning of this call until it eventually fully sounded.

At work, we received notice of a reception at the Officers' Club for Major General Smith, my most important mentor in the 101st Division. He was visiting Fort Rucker. Anyone who served with him in Vietnam was invited to attend the reception. I decided to go, but wondered if he would remember me among the many other people attending. Because Smith was highly respected by everyone, there was a big turnout. He was talking to other aviators when he called out, "Major Boland, you are just who I need for a special assignment in the 1st Cavalry Division."

Then, he turned to my supervising colonel and said, "Surely you wouldn't mind releasing Major Boland for our important mission at Fort Hood."

With little choice of a response, the colonel replied, "No sir, I would be proud to see him have the opportunity."

Smith explained that he had provisionally formed an Aerial Rocket Artillery Battery. He urgently needed someone with my experience to oversee its training. While I was flattered by his announcement, he didn't ask if I was interested in the position. I doubted he was serious, but it was bad thinking on my part. He was serious.

The next morning, I received a call from an artillery assignment officer in Washington D.C. The officer said they had received a by-name request from a general officer for me to be reassigned to the 1st Cavalry Division. I had no time to consider this seriously and was at a loss for words. I indicated to the officer that the timing was bad. I wasn't crazy about moving my kids while they were in school. Thinking this would be a delay tactic, I asked how soon I needed to report. He said, "Yesterday." I shared that I didn't see how I could drop everything and assume this new position. Because it was so seldom they received this type of request, he informed me that it would not be a good idea to turn it down. This special request was also a unique opportunity to strengthen my resume for future advancement.

In today's terms, this decision would be called a no-brainer. Instead, it was time for another important family conference and decision. Jackie loved our home in Alabama. She hated to leave it, but also knew that moving was inevitable before long. Our final decision included remembering and accepting once again that home is where the Army sends you. I needed to proceed soon. Jackie assured me she could manage the move. She stayed behind to prepare to move and sell the house. The decision was made. If we wanted to remain in the military for the long haul, our only decision was to move. And now.

## Fort Hood, Texas

I drove from Alabama to Texas in my very tired 1964 Volkswagen Beetle, which often refused to start. In my new assignment, I was the commanding officer of Company A, 227th Combat Aviation Battalion, 1st Cavalry Division. The F Battery, or the provisional aerial rocket artillery (ARA) unit, was attached to A company. The F Battery commander reported to me. These kinds of arrangements were common in the Division due to test mission requirements. New concepts and innovation were everywhere. It seemed like no idea was off the table.

Although the Army was still fighting in Vietnam, Fort Hood had become the proving ground for the Army of the future. During the years of involvement in Vietnam, the U.S. military adapted a guerrilla warfare strategy for that environment. During this same period, the Soviet Union fielded massive numbers of heavy armored forces. These posed a grave threat to our North Atlantic Treaty Organization (NATO) allies in Europe. Because of our involvement in Vietnam, the U.S. Army had fallen significantly behind in having the capability to confront the Soviet Union threat successfully.

The 1st Cavalry Division, also known as "The Cav" or "First Team," was redeployed from Vietnam to Fort Hood. It was reorganized into a Triple Capability (TRICAP) configuration. The mission was to test new concepts and tactics using a mixture of tank, infantry, and air cavalry formations. The 227th's primary mission was to provide airlift support to the infantry brigade while the F Battery provided backup support to the traditional artillery battalions.

Fort Hood was a sprawling installation with multiple field maneuver areas and live fire ranges. It was an ideal setting for the TRICAP mission. Innovation was everywhere. The Cav even had a motorcycle scout platoon. They also had a policy of distinctive headgear worn by each unit. The cavalry wore traditional black Stetson hats, the artillery wore red berets, and the aviators wore blue baseball caps. These non-conforming uniform modifications were not warmly received by units outside The Cav. The 1972 1st Cavalry Division was unique in many ways and quite different from the rest of the Army.

The cavalry spirit carried forward from Vietnam. Almost all the aviators had served there. Most of us were still in the initial stages of PTSD recovery, but we refused to believe we had any problem. We also didn't realize the significance and favorable advantage of having a mutual support system that we shared with each other. This was in stark contrast to those who left the service and returned to their hometowns to face the problem alone. The black Stetson hat guys referred to themselves as the "Real Cav."

Their slogan was, "If you ain't Cav, you ain't shit." They often shouted this slogan at lively happy hour gatherings on Fridays, which often prompted confrontation with the rest of us. It was a post-war era of high energy and emotion. Our environment afforded these unusual forms of therapy.

When I assumed command of A Company, it became apparent my mission would be much different than I had expected. The company was charged with the responsibility of operating a training school for all aviators in the division so they could receive a standard flight instrument rating. During the period of urgent need to send replacement aviators to Vietnam, instrument training was minimal. The standard rating qualified an aviator to fly under actual instrument weather conditions. This process was conducted locally throughout the Army rather than sending everyone back to training centers. To accomplish this mission, A company was assigned a total of 28 UH-1H Huey helicopters and all of the instrument instructor pilots and instrument examiners in the division. They were all highly qualified and experienced.

The most challenging task was to keep up with the scheduled maintenance for this number of aircraft. A major inspection was required after 100 hours of flying, which often happened. It was common for an aircraft to daily log six to eight hours of flying. In Vietnam, I faced the daunting challenge of trying to keep up with Cobra's maintenance. This assignment may have been more difficult. At least at Fort Hood, Tex., we didn't receive incoming rocket attacks.

Three months into this new experience, another bugle called, announcing the arrival of a new battalion commander. Lt. Col. Frank Henry became my most influential mentor. Again, this was another omen, as I can find no other explanation.

Henry was an amazing leader who exemplified the power of positive thinking. Nothing ever rattled him. He always knew what to do and instilled confidence in everyone around him. Under his leadership style, our battalion excelled in every way. Henry and his wife, Emma, took a personal interest in my family and we became good friends. He encouraged me by praising the performance of my company and the job I was doing. In time, he also recognized my full potential in the Army, and urged me to submit a new application for a regular Army appointment. As part of my application, he provided a glowing endorsement. Major General Smith also provided a similar reference. By the time my application went forward, I was convinced I would finally receive approval.

**Tactical Training Turns Turbulent**

Three months after Henry arrived, the battalion became increasingly engaged in tactical training to support TRICAP testing. All helicopter flights were conducted at nap of the earth (NOE) altitude. This meant that 1st Cav aircraft were not permitted to fly higher than 50 feet above the ground at any time unless it was an instrument flight. The challenges of doing this multiplied when the majority of training took place at night. This was before night vision goggles were available. In fact, no night vision aids were available. Minimum aircraft lighting was required. We accomplished this by covering the bottom half of the small navigation lights with tape. The mission's goal was to fly without detection, which included flying low enough to defeat radar detection, and during darkness hours to defeat visual detection. Navigation lights could not be seen from the ground.

Flying night missions in this manner was challenging. In formations of up to 16 aircraft, the only way to see other aircraft was the upper navigation lights in front of you. Because of the growing number of night missions, Henry deemed it necessary for the A company to conduct night missions, in addition to the daytime instrument instruction flights.

These night tactics were nothing new for Henry. At an earlier assignment, he was part of the 11th Air Assault Division at Fort Benning, Ga., where similar training and testing was conducted. When the division was redesignated as the First Cavalry Division and deployed to Vietnam in 1965, Henry was the Aide de Camp to the commanding general. In Vietnam, Major Henry was assigned to an infantry battalion as its executive officer. The book, *We Were Soldiers Once and Young*, includes details of his heroics during the battle in the Ia Drang Valley. Most of this story centers on Landing Zone X-ray, but a similar fight took place at Landing Zone Albany.

When the A Company joined the night operations, most missions required a 16-ship formation. There were a few occasions when we flew from dusk until dawn without shutting down. Refueling was accomplished at forward area tactical locations with the aircraft still running, known as hot refueling. We also had a special daytime mission, basically a demonstration for visiting officials. This involved a 16-ship formation flown at the nap of the earth to a landing zone in front of a hilltop viewing area. Called the Bushy Knob exercise, aircraft were not visible until they suddenly appeared in the landing zone and unloaded an infantry company. It was a very impressive example of the capabilities of the nap of the earth.

In April 1973, a major force-on-force exercise was held with the 1st Cav. In simulated opposition to the 2nd Armored Division also stationed at Fort Hood, aircraft were flown during daylight with every effort to maintain tactical airspace management at all times. We used tactical boundaries drawn on our maps to keep each division's missions separate. The 2nd Armored Division had a small number of aircraft. In addition to all the simulated combat missions, we were also on standby to perform the Bushy Knob as part of an expected visit of the Secretary of Defense.

Midway through the one-week exercise, a severe storm entered the area. The entire exercise was halted. All aircraft were grounded in place. Just before this happened, I led a mission with six aircraft to place an anti-tank platoon on a hilltop overlooking a potential enemy location. It was very close to the boundary that separated the opposing forces. The Bushy Knob demonstration was placed on hold until the storm passed through. Approximately an hour later, the skies suddenly cleared and visibility became unlimited.

All units were notified to resume full operation. Our first mission was to extract the platoon we had placed on the hilltop. To accomplish this mission, I led a flight of six UH-1H Huey aircraft to do this. The landing zone was small. Only two aircraft could land at a time. Aircraft were spaced to allow enough time for each lead aircraft to land, load the troops, and take off. The second aircraft in each increment flew slightly to the rear of the lead to facilitate a sharp left banking turn after the pickup. We flew at slightly less than 50 feet above the ground and saw no other aircraft as we approached the hill. Our timing was perfect. The last two UH-1H Hueys touched down immediately after the previous two departed.

As I headed back towards the drop off point, I heard a sudden emergency radio call of, "Mayday. Mayday. Midair." I initiated a radio check with my flight and heard them respond, "2,3,4,5…...". My heart sank and my throat tightened as I repeated the call and received the same result no response from aircraft number 6. I had a feeling of terror as I made a 180-degree turn and instructed the remaining aircraft to continue to the drop-off location and stand by.

It didn't take long to see smoke rising at the crash site. When I landed at the site, my worst fears were confirmed. It was our aircraft, inverted and flattened. I could see the crushed pilots still in their seats. Nearby was the wreckage of a smaller observation helicopter but I couldn't tell what unit it belonged to. A medic from a nearby ground unit arrived on foot and assessed the situation. There were ten dead and three survivors. One survivor, the crew chief of our aircraft, was in critical condition. The medic said his heartbeat was intermittent and he had to get to the hospital immediately. While air med-evac had been called, there was no time to wait. Using a broken cargo door from the crashed UH-1H Huey as a litter, we got him on my aircraft and raced toward the hospital at Fort Hood. Waiting hospital staff rushed him to the emergency room.

We headed back toward the crash site. The med-evac ship had arrived and was treating the other two survivors. About this time, I received a radio call from Henry asking me to meet him as soon as I was able. He wanted a briefing on the situation. When I arrived at his location, Division Commander General Shoemaker was with him. As we talked, I still had no idea where the other aircraft came from or to whom it belonged. A check with all 1st Cav units determined it was not ours. It had to belong to the 2nd Armored Division.

Quickly, we figured out what happened. Eventually, a full investigation confirmed our initial conclusion. When the harsh weather moved in, the OH-58 observation aircraft was forced to land at the base of the hill on the opposite side of our flight path. We had no idea it was there. As the weather cleared, they were unaware of our hilltop activity as they took off. As our number six aircraft lifted, it entered a steep bank and turned left. The OH-58 was climbing in a banking right turn. The two aircraft never saw each other, and it was a terrible tragedy. Rapid broadcasting of the accident by local media struck fear among aviator families throughout Fort Hood.

When I finished the discussion with Henry, he informed me the demonstration mission was a go. We had less than an hour to take off. I told him I didn't know if we were capable of pulling it off after what had happened. Personally, I was in bad shape and barely functioning. He reminded me when something like this happened, it was best to get the unit involved back in the air as soon as possible and help overcome the shock factor. He was right, but now I had to face the other 30 pilots awaiting my return and break the news. There were a few grumbles, but they were all combat veterans who had previously dealt with life and death.

We performed the mission.

My next task was to return to Fort Hood and continue to fulfill my duties as the unit commander. This included next of kin notifications and seeing my crew chief at the hospital. I went home to get cleaned up. Jackie was waiting with Henry's wife, ready to assist. When I went into the bedroom, the tragedy's full impact hit me. I started shaking and felt like I was falling apart. Sobbing, I sank to my knees. It was the worst feeling I ever had in my life and the memory of it has stayed with me. It was exactly three years since tragedy struck my unit in Vietnam. Finally, the full accumulation of stress caught up with me.

By the time I arrived at the hospital that evening, I was running on empty. Yet, I had to keep functioning. Jackie was with me. As we entered the waiting room, I was immediately confronted by the surviving soldier's wife. She was understandably still in shock and blamed me for what happened. I wasn't expecting this, nor did I realize this was a common expression of the initial grief stages. Her reaction reminded me that as the soldier's commanding officer, I was responsible for everything that happened in my unit. In this situation, I led the mission that resulted in his imminent death. I could offer no defense. The doctors said they didn't expect him to make it through the night. If he survived, they feared significant brain damage with unpredictable long-term consequences. I didn't sleep much that night and many more nights that followed.

The next day, the doctors were amazed he was still alive. He was in a one-day-at-a-time situation with his chances of survival increasing with each passing day. He was evacuated to the Army hospital at Fort Sam Houston, Tex., for long-term treatment. My yearlong command ended one month later. My replacement and I went to see him. He was still in a coma. There remained no way to know if and when he would come out of it. The new commander kept me posted. I remained in the battalion for another year serving as the S-3 operations officer. Several months later, the soldier began to wake up, but his future was still unknown.

Several years later, I received a letter from him. After many long months of rehabilitation, he gradually restored his functions. He went back to school and began a new career. He had recovered 90 percent and continued to improve. When he finally learned the details of what happened, he thanked me for my efforts to save his life. He expressed a new sense of confidence in himself with no regrets about serving his country as a helicopter crew chief.

After the accident, the Army provided an airplane so a group of company officers and I could attend the two pilot's funerals. These were held in Portland Ore., and Seattle, Wash. The families were grateful we attended. One family pursued the investigation to determine why the accident happened. A final note of this catastrophe: both pilots flew in Vietnam, and one had been gravely wounded. After much rehabilitation, he had recently met the medical requirements to return to flight status.

## Not an Easy Officer Course

During the year of commanding A company, I had little time to devote to the F company or ARA battery. Fortunately, the unit had an outstanding captain. When I became the battalion S-3, Henry tasked me to begin close coordination with the Division Artillery Headquarters to accomplish a full transfer of F Battery to Division Artillery Headquarters. Col. Jack Merritt, Division Artillery Commander, was another outstanding leader and mentor. He eventually became a four-star general. Two of his staff officers, Lt. Col. Bill Schnieder and Lt. Col. Don Jones, became three-star generals. I enjoyed collaborating with them and reconnected with them again in future assignments.

Henry departed in July to attend the Army War College. Another outstanding officer replaced him, Lt. Col. Fred Watke, who moved me to the battalion executive officer position a few months later. The mentoring I received in the 101st and 1st Calvary Divisions was invaluable. These wonderful leaders would have been extremely successful in any career they had chosen.

Shortly after Henry departed, I received notification that my regular Army application was disapproved. I assumed this meant the end of my active-duty career with the Army. It was hard to understand why this happened in light of the robust performance evaluations I had received from my recent assignments. I assumed my National Guard status was holding me back. Even with 13 years of active-duty service and combat assignments, I was still a member of the Wisconsin National Guard. From time to time, I received reassignment and promotion orders from them. It was a crazy arrangement. Thank goodness, this arrangement no longer exists.

I began to think about what I would do in the future when more stunning news arrived. Once again, the bugle called. I received orders to attend the Army Command and General Staff College at Fort Leavenworth, Kas. This was my last year of eligibility. After the regular Army disapproval, I assumed my selection for the Army Command and General Staff College would not happen. My roller coaster ride as an officer in the U.S. Army started another climb. When would this end?

In June of 1974, it was time to move. As a family, we had become fond of living in Texas and wondered if we might ever come back. Dan and Dave were becoming good football players. Now, more than ever, they hated to move. Teen years are so difficult for military families. It is challenging for children to make new friends again. Our friends who had daughters said it was even more difficult for them to break into a new friend group. Football made it a bit easier for our boys. Dan had just finished his sophomore year. The coaches said he was on their depth chart as the starting tight end his senior year. Football is a profoundly serious business in Texas and

Oklahoma. I can attest to the accuracy of the movie *Friday Night Lights*.

On the way to Kansas, we took a detour for an extended camping trip to Colorado. With our pop-up camper and pup tent, we had a wonderful journey through the mountains, trout fishing and rock hunting. We repeated this experience the following summer, and the boys always say those were their most enjoyable family vacation experiences. I agree and realize how precious those days were.

After years of stress, time suddenly stood still. Our family was finally together alone. The camping trip was a wonderful opportunity to strengthen our family bond and share our full love and appreciation for each other. In the solitude of the Rocky Mountains, we felt a sense of peace and tranquility that is hard to find in the hectic day-to-day pace of Army living. With no phones or television, we simply enjoyed one another. We played cards and games and sang together. The 1974 music of John Denver and Elton John was on our lips, exactly in the right place to belt out *Rocky Mountain High*. To this day, I occasionally find myself humming *Crocodile Rock*. Some days, we ventured into altitudes above the timber line where we gazed upon the mountain peaks and the sky above them. As I looked upward, I felt as though I was holding hands with heaven. And I didn't want to let go.

# Chapter 13

# Career Soldiers

From the moment we arrived at Ft. Leavenworth, Kan., I could feel my military career was starting to change. It was exactly 13 years since we had driven from Wisconsin to Fort Sill, Okla. So much had happened to my family and me during those years.

For the first time, things felt different for me as a career soldier. The Command and General Staff College (CGSC) class of 1975 was welcomed with open arms. My classmates and I felt a sense of importance for being chosen for the school. Immediately, we were assigned to on-base family housing. It is amazing how no house hunting and school choice selection made this move feel so much easier. Elementary and middle schools were on-base with one senior high school in the city of Leavenworth, Kan. Our four-bedroom apartment was at the end of a four-plex building among a horseshoe of similar buildings. All our neighbors were classmates. Quickly, we assimilated with the other families and got to know each other quite well.

A main takeaway from CGSC was the wide range of talent and experience among the class members. There were over 1,000 students. Every branch of Army service was represented along with a small number of soldiers from the U.S. Airforce, Navy, and Marines. Most of the Army students had the rank of major. There were nearly 100 students representing 57 different allied countries. Just one year after the 1973 Arab-Israeli war, some students shared first-hand opinions from that region.

The class was divided into 23 sections or study groups. Each section was assigned a full mixture of different branch experiences. I was assigned to section 14. Our section leader was a lieutenant colonel from the staff judge advocate (JAG) branch and a lawyer. Another lieutenant colonel assigned to my section was an ordinance officer. My tablemate was a military intelligence officer. The daily interaction with top-notch officers from every Army branch was an invaluable experience. No matter what staff planning exercise we were assigned, we always had experts from multiple areas to lead discussions.

After spending so much time in combat unit operations, it was a refreshing revelation to be surrounded by many diverse points of view. The course curriculum was rigorous and challenging with excellent instructors. I learned so much more than I anticipated. When I graduated, I felt ready to serve as a battalion commander or at any level of staff responsibility.

The 1974-75 school year was unique because of the major changes taking place within the Army. This was not foreign to me because I had participated in testing new concepts at Fort Hood. The difference was how everything encompassed a much broader scale. It seemed every Army field manual was under revision, which meant our study materials were subject to change on a daily basis. The Army Training and Doctrine Command (TRADOC) was leading an Army-wide doctrinal revolution. CGSC was a vital component in making sure officers knew the changes.

## The Final Fall of Vietnam

While at CGSC, we experienced the sudden, dramatic and unexpected fall of Saigon, Vietnam, on April 30, 1975. Most students had served in Vietnam. The stunning realization that everything we and our comrades had fought for and sacrificed was lost. My fellow officers and I felt a huge letdown. The U.S. had invested so much into the region. Suddenly, it was over.

The book and briefings we received at the college suggested the South Vietnamese could have successfully defeated the invaders. Catastrophic decisions made by their national command authority prevented this. When soldiers were told to make a full retreat to the coastline and defend the cities, a state of panic completely clogged all the major roads. No one could move. Complete disaster followed.

This was a deeply emotional moment for my classmates and me. Four of our classmates were Vietnamese officers in an army that no longer existed. It was terrible to witness their heartache and disappointment. A merciful decision was made to allow these officers to complete the remainder of the school year and participate in our graduation ceremony. After graduation, they were on the street looking for work, unable to return to the place they called home. I've often wondered what happened to them.

Foremost in my mind was my friend Major Tuan. He had been serving as an artillery commander in the central highlands near Pleiku. In college briefings, we learned the Pleiku area was a focal point for the North Vietnamese invasion. Many South Vietnamese soldiers were killed or captured during the fighting. I couldn't stop thinking how Major Tuan had predicted this outcome six years earlier during dinner at my home in Fort Sill, Okla.

After Saigon fell, Major Tuan was taken prisoner and held in a re-education camp for many years. Eventually released, he only had his spirit and dream left. Literally, he began anew with his bare hands repairing bicycles in the streets of Saigon. His story is detailed in the book, *Goodnight Saigon*.

In 1995, I saw a television news story about the 20th anniversary of Saigon's fall. I was stunned when my friend was interviewed and spoke about the successful factory he established in

Saigon. My attempts to contact him were unsuccessful. Unfortunately, he passed away shortly before I traveled back to Vietnam. His dream remains alive and well as part of a booming capitalist economy that is among the fastest growing in the world. I hope I live long enough to see the ultimate fulfillment of his prediction. I am convinced it will happen someday.

It took a while for those who served in Vietnam to recover from the stunning sadness that fell upon us. The pain of this tragedy has continued to stay with me. When I was finally able to visit Vietnam in 2018, I gained a more long-term perspective of what the U.S. military accomplished while we were there. I became more convinced than ever that the American spirit of freedom has stayed in Vietnam, just as Major Tuan predicted.

## Settling into Leavenworth

My year of Leavenworth schooling provided lots of wonderful and quality family time. With normal five-day school weeks and no weekends or extra duties, our family relaxed and enjoyed life. Our sons were quite grown now; Dan was 16, Dave 14, and Mike 10. We found ourselves caught up in a big year for football. Both Dan and Dave excelled on their respective teams. After football, Dave had a highly successful wrestling season and was a standout in track and field competitions. Dan started on offense and defense and showed great promise as a defensive end.

Our third son, Mike, was not as big or athletic as his brothers. So far, he had avoided sports, reluctant to try to live up to their success. While in Leavenworth, Mike played soccer and enjoyed it a lot. Best of all, he thrived with competition and teamwork.

Our family became particularly good friends with our next-door neighbors, the Kwieciaks. Stan was also a major and field artillery officer. With his wife Renata and two young sons, they were a genuinely nice family. Later, Stan was promoted to general officer. I was immensely proud of him. Several other classmates earned the general officer rank. They were each very deserving.

## Life After Leavenworth

During our schooling, two topics that garnered a lot of discussion were Officer Efficiency Reports (OER) and Orders. Official orders were published for everything, such as awards and promotions. Another favorite discussion topic revolved around our next assignment. We were all vested in where each of us would move and what our next assignment would entail. Some students considered a Pentagon assignment as a prestigious next step. Serving there as a major action officer was considered an important credential for bigger and better career advancement. Anticipated orders were a source of great anxiety for family members. We all wondered, "Where will we go from here?"

Unlike my fellow students, I was still very naïve about getting ahead as an officer. I still fell victim to the idea that if a soldier did a decent job in whatever assignment they had, the system would care for them. I failed to recognize that I was in a different league now. From this point forward, career advancement was an incredibly competitive process.

Each officer needed to set their own goals and actively seek opportunities to achieve them. I was still riding a wave of gratitude that I'd even been selected to attend the CGSC and wasn't thinking about the best option for my next assignment. My goal was simply to be promoted to lieutenant colonel and complete 20 years of active-duty. Nothing beyond this was on my radar screen. Some officers worried too much about where they were headed and how to get there. An important piece of advice I learned while serving in the 1st Cavalry Division was shared in the Division Commander's philosophy of, "Don't take yourself too seriously." While I tried to maintain this perspective, my career aspirations soon started to grow.

Assignment officers from each of the Washington D.C. Army branch offices came to Fort Leavenworth. They met with us and discussed our next assignments. Lt. Col. Don Jones was the field artillery representative. I met him while at Fort Hood and remembered him as a straight shooter. I was confident he would give me good advice. I was not disappointed.

He talked honestly about my long-range potential and the importance of being competitive with other field artillery officers inline for promotion to lieutenant colonel. Jones pointed out that my extended time in aviation assignments meant my field artillery credentials were weak. While he commended my strong aviation assignment evaluation reports, he noted that I was competing with officers who had similarly strong records in field artillery assignments.

Wisely, he recommended the best assignment for me would be the field artillery school at Fort Sill, Okla. There, I could immerse myself in the mainstream of evolving field artillery doctrine, become a legitimate artillery expert and be fully branch-qualified. Grateful for his counsel, I never saw him again. I followed his rise to the rank of three-star general with admiration. While my family would have loved to stay at Fort Leavenworth, it wasn't in the cards.

## Fort Sill - Round Four

Prior to graduation at Fort Leavenworth, representatives from the communities we were moving to visited us. Of particular interest were real estate discussions. My D.C. bound classmates were swamped by agents hoping to sell them a Washington D.C. area home. There was absolutely no hope for a major to get on-base military housing. A representative from the Fort Sill area met with us and we learned about the community. The representative's husband was with her. We wound up talking more about football than housing. They had a son who was entering his senior year and a starter on the Lawton Eisenhower High School football team.

Our conversation turned into a recruiting effort for Dan and Dave to attend Eisenhower, which is where they attended. We bought a home in this school district area and later moved on-base when housing became available. There were three high schools in Lawton, OK. The cross-town rivalries were intense.

For the next three years, football became a major family focus. Eisenhower played in a conference with the largest schools in Oklahoma, so we traveled all over the state. Dan's team began the season as a ranked contender but had a disappointing season. During his senior year, Dave's team went undefeated until the semifinal game for the state championship. Twenty thousand fans attended the rivalry game with Lawton High School. The whole experience was unlike anything we had known.

Back at Fort Sill, I felt an intense sense of nostalgia. This was the very place I served as a private nearly 20 years earlier. As I pondered my new opportunity to become a field artillery expert, I recalled my 1956 days of being a cleaning toilet expert, which felt like a lifetime ago. My return in 1961 as a second lieutenant and in 1968 as a major seemed like distant memories. Now, I was back in Fort Sill for my last hurrah. I never expected another unsuspecting bugle call with more surprise events.

I was assigned to the field artillery school in the Target Acquisition Department. This department was not considered the most exciting place to work. I wondered how I would ever acquire the credentials Jones referenced. The Gunnery Department and Tactics Department were considered much more important. I wondered why I wasn't assigned to one of those departments. Little did I know that target acquisition was about to explode (no pun intended) as the most important center of new doctrine in the entire field artillery school. Soon, a huge bugle call sounded. I became a primary figure in the development of a new concept called counterfire.

The post-Vietnam era brought with it a frenzy of innovative ideas throughout the Army to prepare for future conflicts. My personal involvement with revamping Army doctrine began at Fort Hood, continued in my studies at Fort Leavenworth, and spilled over into my assignment at Fort Sill. My progression through these steps gave me a distinct advantage in seeing the big picture and visualizing the role of field artillery within it. Honestly, I was in the right place at the right time. It became clear that I could have a significant impact on the Army's new mission.

Counterfire became the new name of the Target Acquisition Department. The dynamics of this new concept suddenly made the department an exciting place to work. Counterfire's definition of "the attack of the enemy's indirect fire system" became an aggressive concept. Ultimately, the aim is to defeat every aspect of enemy field artillery capabilities; not just weapons.

Important advancements in technology produced new radar systems that greatly increased our ability to simultaneously locate multiple enemy gun locations and attack them. New attention was placed on enemy communications systems as well as re-supply capabilities. The far- reaching impact of these concepts involved revision and rewriting of new procedures and tactics to implement counterfire throughout the Army.

I oversaw the work needed to accomplish much of this. Eventually, we were directed to assemble a team to brief the entire field artillery community about this new doctrine. Training sessions were held at multiple locations in the U.S. as well as in Germany and Korea. The entire process took place over a two-year period and spilled into 1978. During this time, I had the opportunity to interact with senior leadership throughout the Field Artillery Division. I can't say my name became a household word, but I met a lot of folks that I would have never known otherwise. My central role in this new initiative placed me at the forefront of field artillery employment doctrine for the modern battlefield.

I was promoted to lieutenant colonel in November 1976. This promotion placed me into the assignment where I became responsible for everything just described. It was an extremely proud moment to have the silver leaf insignia pinned on my collar. My family shared the day's happiness. We felt a keen sense of having reached this milestone together. I would have never achieved this goal without the selfless sacrifice and unwavering support from Jackie. How the boys lovingly supported their mother and endured their own hardships and disappointments made this day the culmination of a remarkable team effort. Truly, the unsung heroes of military life are the families who sacrifice much. Enough praise can never be bestowed upon them.

The promotion sealed my tenure status to complete 20 years of active-duty and receive long-term retirement benefits. I still had a few years remaining until my mandatory retirement date of July 1981. Because three years was the normal maximum length for any assignment, it appeared I would have one more assignment. In fact, my current assignment was the only one that had lasted the entire three years.

The nature of this assignment wasn't a huge concern. Later, I started to think more about life after the Army. I wasn't anticipating another important bugle call. But at this point, I should have realized that throughout life, another bugle can always sound.

The three years at Fort Sill were extremely rewarding in many ways. We became close friends with several of my co-worker's families. We shared a love of boating, playing cards, singing, and the famous aroma of Ray's homemade pizza. On many Sundays, our core group combined all of these activities together. Two wives played the piano. Often, we conducted choir practice after pizza and cards. The choir was capable of singing off-key horribly but their extraordinary enthusiasm made up for this shortcoming.

Looking back, I have fond memories of these years. This was the best period of family activity we ever enjoyed. In 1976, Dan graduated from high school and entered college at Cameron University in Lawton, Okla. He lived at home, joined ROTC, and did very well. At the end of his first year, he was offered an ROTC scholarship. A year earlier, he had been offered an appointment to the West Point Prep School where he could have played football and strengthened his academic standing. At the last minute, he turned it down due to uncertainty about the military obligation if he graduated from West Point. Looking back, he answered the best bugle call for his future.

As my time at Fort Sill wound down in 1978, I received a most unexpected phone call from an assignment officer at the Field Artillery Branch in Washington D.C. His first words were, "Ray, we have some great news for you."

Hesitatingly, I said “What?”

Then, he spoke the most unbelievable words I could hear. He said, “You’re going to be a battalion commander.”

For a moment, I was speechless. Finally, I squeaked out, “How can that be?”

He explained that I had been on the unpublished alternate Command Selection list. An unexpected command position had opened up. I was next on the list to be assigned. Still in disbelief, I asked him to state my social security number, so I confirmed that he had the correct Ray Boland.

I was still struggling for words when he said, “Don’t you want to know where the assignment is?”

I replied, “I guess so.”

I would be commanding the 25th Combat Aviation Battalion Division, part of the 25th Infantry Division in Hawaii. Now, I felt completely blown away and couldn’t imagine how this was happening. For years, I submitted assignment preference statements, known internally as dream sheets. Every time, I listed Hawaii as my top choice. Everyone did this. It was almost a joke. I never knew anyone who was fortunate enough to be sent there.

Before this conversation ended, the assignment officer said, “By the way, we want you to submit a new application for a regular

Army appointment because your situation has changed."

All of this new assignment seemed to fly in the face of the goal to strengthen my field artillery standing. I was now returning to the world of Army aviation.

And that was that. I had just received a bugle call for the ages. In a few short years of Army service, I went through a classic rags-to-riches transformation. The stage was set for a new beginning of my Army career and service to the nation. My state of uncertainty had shifted from whether I would last for twenty years to how far I might be able to go in thirty years. I was completely dumbfounded. How could a boy who grew-up on the south side of Chicago now become a battalion commander? And be stationed in Hawaii?

I recalled and remembered the small seed that I felt planted in my heart the day the Japanese bombed Pearl Harbor. I was now going to the same area that had been affected on the day I first became aware a U.S. military existed. This time, my heart was not filled with the angst my mother was overwrought with. This time, I felt great peace and comfort in knowing that my family and I would be assigned to one of the best military opportunities in the U.S.

For a moment, all the moves, challenges and overseas trauma felt a little serendipitous. I had taken to heart President Kennedy's call to do something for my country. A little reward for the years of service, sacrifice and scrimping for our family now seemed to be firmly planted into my outstretched and cupped hands. I opened my head and heart to receive this assignment, which felt like a huge gift. I could only thank the Lord, my wife, my sons and all those who had helped me mature into the man and soldier I had become.

I was determined not to let anyone down. I would treat this present with the utmost care and continue to serve my country to the best of my ability. The Army had invested so much in me. It was now my time to give back to those who had guided me through multiple bugle calls, as well as those I would now be responsible for.

A celebration followed with all my colleagues and friends and many well-wishers. I felt like I was living on the set of an *Everyone Loves Raymond* episode. Sometimes, dreams do come true.

# Chapter 14

# Tropic Lighting

After the initial excitement of my assignment to Hawaii calmed down, I was able to take stock of the full meaning of this amazing bugle call. I felt a sense of honor to serve in another combat division with a proud history. Known as the Tropic Lightning Division, the 25th Combat Division served in World War II, Korea, and Vietnam. It was an important rapid deployment force for contingencies in the Pacific Theater.

The 25th Combat Aviation Battalion was considered the largest of its kind in the Army, although a friend commanding a similar-sized battalion in Alaska claimed he was slightly larger. With nearly 100 helicopters and more than 1,000 people, commanding this battalion was a big responsibility. Established as a separate battalion and not associated with a brigade headquarters, the commander also served as the Division Aviation Officer, which was a special staff position. Additional assigned aircraft were in the Division Cavalry Squadron.

What I looked forward to the most was the privilege of serving with the men and women assigned to the battalion. I was committed to providing my absolute best leadership. In each of my three previous company command assignments, I learned soldiers are the heart and soul of everything that happens. Mindful of my own experiences as an enlisted soldier earlier in my career, I remembered the popular saying, "Shit runs downhill." The troops at the bottom usually suffered the most from bad decisions made at the top. I tried to remind myself each day of this mantra.

Maintaining a consistent leadership style was my goal. Being undeviating is essential. Another popular Army saying was, "You can't fool the troops." Regardless of rank, soldiers are very savvy. They can spot a phony in a hurry. Soldiers can tell when a leader's actions are self-serving rather than what is best for the entire unit. Soldiers expect leaders to enforce standards consistently. Failure to do this is considered a sign of weakness. Using my own military experience to understand the significance of these attitudes was extremely helpful as I strove to be effective commander.

Preparing for this new adventure included some important family decisions. Families are significantly affected by certain events of the soldier, especially when it comes time to move. This was another one of those times. Our sons were excited about the possibility of living in Hawaii. They saw this as a once-in-a-lifetime opportunity. Dan decided to continue his college studies at the University of Hawaii. Dave had just graduated from high school and was interested in attending a Hawaiian community college. Mike was entering high school. It didn't take long for everybody to decide they were all in for living a life in paradise.

A more difficult decision was what to do about our beloved family puppy, J.J. A lovable farm mutt we acquired five years earlier in Wisconsin, she had become an incredible family companion. Our chief concern was the extended quarantine period required of all pets entering Hawaii. How would she deal with this? We developed a close relationship with another Army family in Oklahoma. They loved J.J. as much as we did and offered to care for her while we were stationed in Hawaii. They became so attached to J.J. that she remained with them the rest of her life. By now, camping was outdated for our family. Boating had taken its place, and we had a wonderful water-skiing boat. We towed it to California and paid to ship it to Hawaii.

As we sorted through these decisions, I became aware of various pre-command orientation and training requirements at various locations. This meant an extensive period of temporary duty before going to Hawaii. This series of requirements helped me become current in a wide range of important military subjects. The lengthiest training involved a trip back to Fort Rucker for a helicopter refresher training course and renewal of my instrument flight qualifications.

Prior to departing from Fort Sill, I received some disturbing news. I met with an old friend who served in the 25th Division. He informed me of a potential problem with my assignment. The division commander had notified Washington D.C. in writing that he was rejecting my assignment to command the Aviation Battalion. He felt the Army should not waste this command opportunity on a National Guard officer who had so little time remaining in service before facing mandatory retirement in 1981. Given the policies at the time, this was actually a reasonable point of view. For me, it was a knockout punch. My friend advised me to remain calm and let the situation work itself out. Easy for him to say!

I wasted no time and contacted the Field Artillery Branch assignments office in Washington D.C. They confirmed receipt of the letter from the 25th Division Commander. Naturally, I asked, "What happens now?"

I was pleasantly surprised to hear the branch office was standing firm on my assignment and had informed the division commander there was no basis to change my assignment. They were following Army policy for command selection. I was the next person on the list. They informed the division commander that I had applied for a regular Army (RA) appointment which would extend my active-duty eligibility up to 30 years.

The general responded, "What if he's not approved?"

The powers that be said I could reapply.

Hearing about this unusual communication exchange helped

me realize the Army was looking out for me and they weren't backing down. What a relief!

My Hawaii report date was scheduled for late August. The assumption of command ceremony was planned for early September. When I arrived at Fort Rucker in late July, I was greeted with another piece of troubling news. As I signed in, I was told to contact the Washington D.C. Army headquarters immediately. I asked for additional information, which could not be provided. I wondered, "Now what?"

I was provided a point of contact (POC) in the Office of Congressional Legislative Liaison (OCLL). I did not recall hearing of this office before and started to become anxious. By this time of my life, I had learned how things can take a negative turn when least expected. Sometimes, this happens just when things seem to be going very well. This felt like one of those times.

The point of contact informed me that I needed to go to Washington D.C. the next day so I could testify before the Senate Armed Services Committee. My question was, "WHY???"

My name was on a regular Army appointment list that required confirmation by the Senate. The Senate Armed Services Committee had a long-standing concern about rubber stamping selection lists with no contact with any selectees. This time, the committee had put their foot down and refused to sign the appointment list until the Army arranged for officers on the list to appear in person before the committee. Another officer on the list and I had been randomly selected to appear.

I was shocked. Why was this happening to me? I confirmed that the Army could count me in. Unfortunately, they had no information about what the subject of the testimony might be. "Swell!" I thought to myself. Why is there no information about the testimony?

Once in Washington D.C., I was reminded that no one was aware of what the committee might ask. I was also instructed to avoid expressing an Army position on any subject. Instead, I was coached to state my own opinion. In other words, I was on my own. "Let your conscience be your guide," was the best recommendation I received.

My thoughts? Thanks a lot.

I was asked if there was an Army staff person that I wished to speak to before I testified. I thought this sounded odd, almost like a last meal request. Fortunately, I knew the two-star general in charge of OCLL. I served under him at Fort Hood. My request was granted. He helped calm my concerns and put me at ease. He assured me there was nothing personal to be concerned with and encouraged me to answer the questions as directly as possible. The staff didn't anticipate anything problematic with my assignment.

The hearing was more relaxed than I expected. The chairperson was very gracious and thanked us for taking the time to appear before the committee. He assured us that nothing we said would be held against us. If anyone in the Army questioned our testimony, we were to contact him personally. While this was a nice gesture on his part, I knew this was not a realistic scenario. If the Army was unhappy with my testimony, this would not bode well for me.

To my surprise, a primary focus of the questions centered around women in the Army. By 1978, the Army had opened up many of the previously male-only skills to women. The committee asked my opinion on how this was going and what problems existed. I talked about the significant cultural adjustment taking place and I indicated it would take time for this to be resolved. At the time, I shared my opinion that the sudden infusion of women into an all-male culture was something soldiers weren't ready for and didn't know how to deal with. Issues ranged from resentment to paternalism, as well as sexual harassment and abuse.

While I am not directly involved in soldier life today, these are things we still hear about. In 1978, I did not expect that some of these same issues would still exist in 2023.

The Committee seemed satisfied with my testimony. They thanked me for being honest. My point of contact indicated that I did okay.

After the hearing, I was taken to see a three-star General for a debriefing. During this meeting, I received the impression the Army anticipated a much different direction during the hearing. I also began to suspect that my selection may not have been as random as I was told. I never had reason or evidence to confirm this. Most importantly, I was satisfied and relieved to have the whole experience behind me.

The good news? The Senate signed off on the regular Army list and I was approved. I exhaled a quiet, "Amen." I returned to Fort Rucker to complete my flight training. Finally, I was ready to proceed to Hawaii, where the division commander still did not want me.

## Welcome To Paradise

From the moment our family landed at the Honolulu airport, the ever-present tropical flowers fragrance seemed to swallow our hearts. Each time I have arrived in Hawaii, this experience has been the same. Hawaiian leis are sold in airport terminals and are an automatic part of the arrival ritual.

The immediate feeling of Hawaiian paradise is one that never goes away. There is also a feeling of being in another country rather than the U.S. It's something one must experience firsthand to fully appreciate.

I was fortunate to succeed Lieutenant Colonel Terry Henry as Commander of the 25th Aviation Battalion. Henry and I had been flight school classmates. He was a fellow field artillery officer as well as a friend. An exceptional person and an outstanding commander, Henry was designated as my sponsor to coordinate arrival and transition into the command position. His wife, Kay, picked us up at the airport and took us to a hotel. Their hospitality was excellent. The boys were in awe of their new surroundings and couldn't wait to test the Waikiki surf. We were authorized up to two weeks temporary living allowance at the Hale Koa Army Hotel on Waikiki Beach, a perfect start to a fantastic experience for us.

Quickly, I learned that the 25th division commander was concluding his period of command before my assignment began. I never met him. My change of command ceremony took place on September 8. It was a typical sunny Hawaiian day. The ceremony was attended by all the Division senior leaders and their spouses. It was a very impressive event, capped off with the Division marching band. In a fitting welcome for a new commander and spouse, Jackie received so many leis from other spouses that her huge floral necklace was almost suffocating.

The ceremony was on a Friday. My newly-acquired staff said they needed to give me a short briefing before the weekend began. We adjourned to the headquarters. The battalion headquarters and most of the battalion units and aircraft were located on Wheeler Air Force Base, which is adjacent to the Schofield Barracks Army Base. Wheeler Air Force Base was home to Army Air Corps units when the Japanese attacked Pearl Harbor in 1941. Most of the U.S. aircraft were destroyed or damaged during the attack before subsequent attacks upon Hickam Air Base and the Naval Fleet in Pearl Harbor. Damage was still evident to the original hangar buildings. I quickly realized this was truly an important historic place to live and work.

The briefing covered the deployment of Cobra gunships to the uninhabited Island of Kahoolawe for night gunnery training. For decades, this island was used by each of the military services for live firing. The Navy controlled the use of the island and approved requests from the other services for its use. Our unit was the only scheduled user during this period. The training was scheduled to qualify crew members under nighttime conditions. They would deploy on Sunday and begin the training Monday night. Coordination was made with an Air Force unit at Wheeler Air Force Base to illuminate the target area by dropping flares. Based on my own Cobra experience, I foresaw no shortcomings with the plan. I was satisfied this would be a good training operation and wished them well.

## Catastrophe Strikes Again

At 10:30 pm Monday night, I received a phone call from the battalion executive officer. His unforgettable first words were, "Sir, I think we have a problem." An Air Force pilot had just informed the Wheeler control tower that a friendly fire incident happened at the Island of Kahoolawe. Eight Marines were wounded. There had been no direct radio contact with our unit on the island, a significant shortcoming. No other details could be confirmed. Fortunately, the deployment included a medevac aircraft which was reported to have taken the casualties to the nearest hospital in Maui. I requested an aircraft and crew to take me to Kahoolawe as quickly as possible because we definitely had a problem.

At approximately midnight, our UH-1H helicopter departed Oahu. We flew along the bright lights of the Honolulu shoreline, a truly beautiful sight. Moments later, it occurred to me that we were over the dark vast expanse of the Pacific Ocean with no place to land. Exactly then, I realized the increased level of risk my air crews were exposed to every day. This new variable added a whole new dimension to my previous aviation experience. There was plenty of time to think more about this in the future. For now, my focus was to find out how the tragedy had occurred.

Upon arrival, the captain in charge provided me with a detailed briefing. The details were hard to believe and were much more complicated than previous situations I had encountered. The entire unfortunate event was bizarre.

On Monday morning, a Marine helicopter unexpectedly arrived at Kahoolawe with a platoon of Marines. Their officer in charge explained to our officer in charge that they would like to conduct some daytime training with their anti-tank weapons. The Navy told the Marines it was up to our guys to approve their training because the Army had the island reserved. Our Army officer in charge informed the Marines that it was okay to train as long as they were finished by the end of the day. Their helicopter took them to the training area and returned to the helipad parking area to await the end of the day. Our Army personnel expected the Marines would return to Oahu that night.

From this point on, the story becomes crazy. At days end, as expected, the Marine helicopter crew bid farewell to our guys and took off. Their helicopter was heard departing the training area. Our guys logically assumed the Marines had returned to Oahu. But this was not so! The Marine's officer in charge decided to remain on the island overnight and continue training the next day. He asked the helicopter to return the following day to pick them up.

Unfortunately, the Marine's officer in charge made no effort to coordinate with our guys his decision to stay for another day, despite having radio contact with them. He also decided to make their overnight stay a tactical training exercise. The Marine troops dug in and camouflaged their position.

At dusk, our Cobras did a last light range sweep to confirm the training location was clear. The well-camouflaged Marines were not detected. To complicate matters, they had turned their radios off and made no attempt to inform the Cobras they were still in training. Finally, they chose a position adjacent to the intended start-fire line for the Cobras. All of this added up to a disaster in-waiting.

Once darkness set in, the Air Force aircraft arrived and provided the flare illumination. Then, our Cobras lined up to start firing training. The first run was to train a front-seat pilot how to operate nose turret weapons, starting with the 40-millimeter grenade launcher. The bursting effect of these is similar to an infantry hand grenade. The instructor pilot announced, "Commence firing."

Moments after the first firing burst, an emergency radio call came in screaming, "Mayday. Mayday. Cease firing. Cease firing."

The grenades had exploded in the Marine's position. They had multiple casualties. Our medivac helicopter was enroute to the scene quickly and became the true hero of the night.

The medic on the helicopter was combat experienced in Vietnam and knew exactly what to do. He performed a rapid triage of the wounded and determined which were most in need of emergency treatment. One person was considered critical. Quickly, the helicopter was on its way to the Maui hospital, which was only 11 miles away, with three Marines. Because of the medic's quick actions, there were no resulting fatalities. To my knowledge, they all successfully recovered.

After hearing the story from my captain, I had no reason to be upset with him or his crew. I thanked them for their quick reactions. The captain felt terrible about what happened. I assured him that I didn't think he was at fault. I asked him to carefully write down everything that happened. We would prepare a detailed after-action report to prevent something like this from happening again.

After the medivac helicopter radioed that everything was under control in Maui, I headed back to Oahu. Enroute, I received a message from the battalion executive officer that the division commander wanted me to brief him at 8 a.m. I anticipated that he would want my report as soon as possible. I considered that I might be facing the shortest battalion command period in the history of the U.S. Army. This was a profoundly serious incident. Commanders are accountable for everything that happens or fails to happen within their command. I was prepared to face the consequences. My previous experience taught me that during life's worst moments, we have to let our faith take over. There are some things we have no control over. My best course of action was to depend upon a higher authority.

When I met with the division commander, I made no attempt to sugarcoat anything. I said, "Sir, this is a complicated story, but I'll start with the bottom line. Our Cobra pilot pulled the trigger that shot and wounded the Marines."

I had only spoken to my Commanding General once before this meeting – just a few days earlier at the change of command ceremony. We didn't really know each other. In the next few moments and in the weeks and months ahead, I learned that Major General Clyde Lynn was the best division commander for whom I could hope. He was firm but fair, and valued honesty more than anything else. He also valued performance over credentials or background. He didn't care in the slightest that I was a National Guard officer – only what kind of commander I was. After patiently listening to the whole friendly fire incident, he didn't fire me. This was not a great day for him or the 25th Division. The entire situation was a major black eye.

Headlines appeared in newspapers all over the U.S. stating, "Army Shoots Eight Marines." A full joint service investigation was directed. The Navy, Army, Marines, and Air Force all participated. It was a very big deal. I don't recall what the final report included but there were no punitive actions recommended for any Army personnel. For everyone concerned, the most important outcome was the survival and recovery of the Marines.

After such a traumatic beginning for my 18-month command period, the remainder of my assignment could seem barely worth describing, but it was. Personally, this opportunity became the most gratifying experience of my entire Army career. While serving in paradise was part of the lure, collaborating with many outstanding battalion officers and soldiers was the most important reason I deeply appreciated this assignment.

Many believe that battalion command is the most important job in the Army. I cannot disagree. Company commanders are remarkably close to their soldiers and non-commissioned officers on a daily basis. They have significant direct involvement in the unit's performance. Battalion commanders can influence success in a unique way by sharing lessons learned from a wide range of experiences. Maintaining a positive command climate is especially important. My predecessor, Lieutenant Col. Henry, accomplished this. My task was to maintain this culture, not create it, or even worse, screw it up.

Frequently, I thought of the other Lt. Col. Henry I knew. Unrelated to my predecessor, this Lt. Col. Henry was my boss at Fort Hood five years earlier. I didn't try to replicate his leadership style, but I often did recall his decision-making skills in dealing with various situations. These reflections proved extremely helpful.

## Finally – Regular Army

Approximately one month after assuming command, I was officially sworn in as a regular Army (RA) officer. A small ceremony was conducted by my boss, Brigadier General Art Brown. He remarked about the rare circumstances that lead to such an appointment so late in one's career. It's likely this occurred with other officers, but I am not aware of who they were. Few people were aware of the long road traveled to reach this goal, including the Senate testimony. None of this mattered now. I had finally become a member of the A team. I intended to make the best of my remaining service.

While the island of Oahu is a world-famous tourist destination, it was also an important military center. Each branch of our armed services had a significant presence in this relatively small piece of real estate. Most Army units were located at Schofield Barracks, the movie location for *From Here to Eternity*. Pearl Harbor and Hickam Field Air Base are also famed in movies. Perhaps less known is the Kaneohe Marine Base on the eastern side of the island. Combined, these installations make Oahu a significant military fortress and a key center for military response to contingencies throughout the Pacific Theater.

Hawaiian culture is a unique living experience. The hustling, bustling nature of Honolulu made it a fascinating place to live. We had no idea what it was like to live where Caucasians were the minority population. Dan and Dave had employment opportunities at military facilities based on minority quotas. We were surprised to learn of the large Portuguese influence that dates back to colonial times. The wide ethnic population diversity and customs made it feel like we were living in another country. Yet, this diversity made every day special.

Much of our Army training took place on the island of Hawaii rather than Oahu where training areas were limited. This meant thousands of flying hours over the water between the two locations with more than an hour of flying time in each direction. The island of Hawaii presented a completely different terrain situation than Oahu. In addition to active volcanic activity, Hawaii is home to two separate snow-capped mountains: Mauna Loa and Mauna Kea. The airfield training base is 6,000 feet above sea level, compared to lower sea level elevations on Oahu. In this environment, saltwater corrosion is an aircraft maintenance challenge. Each aircraft required constant inspecting and cleaning.

Everyone in the battalion needed to remain vigilant in every aspect of mishap prevention. The potential for a serious mishap was ever-present. I confess that I felt a constant level of daily stress during all of this training and deployment.

An important monthly event was the required preparation and forwarding of the Unit Readiness Report. These reports went to the Department of the Army Headquarters where the combat readiness of all worldwide units was monitored. General Lynn required each battalion commander to brief him on their monthly report. This was a serious show-and-tell event. Commanders weren't permitted to bring along staff officers to assist. The commanding general referred to this as an opportunity for each commander to demonstrate their ability to withstand the heat of the executive session. I actually enjoyed these opportunities and was comfortable presenting the information and answering questions. My ability and confidence to complete this trailed back to the experience I gained through mentoring, all the way back to General Smith during my 1970 tour of duty in Vietnam. I was used to closely following daily unit status and was provided data each day to accomplish this. I didn't have to cram at the end of the month. Each of my subordinate commanders shared this intensive management approach with me so we always had a mutual understanding of our posture.

This arrangement was part of the command climate that I assimilated earlier in my career. It was not micro-management from above. Instead, subordinate commanders knew their boss had their back when the going got tough. This was an environment that cultivated complete honesty and a joint adherence to the motto that,

"Bad news doesn't get better with age."

I shared these same understandings with my boss, Brigadier General Brown. During my entire command, he was an important mentor who made me feel at ease as I performed my job. This relationship continued beyond our service in Hawaii. At the time of our last conversation, he was a four-star General serving at the Pentagon as the vice chief of staff for the Army. The last I heard, he was still going strong beyond the age of 90. He was truly a great guy.

The cultural adjustment process raised during my Senate testimony was in full progress in my battalion. I learned a lot more about the benefits of women in the Army. Two of the Army's first female helicopter pilots were assigned to my battalion. Many female soldiers helped with aircraft maintenance and other duties. I appreciated equal opportunity and feel female soldiers are capable of performing a wide range of duties.

Yet, traditional ground rules for fraternization grew fragile and challenged. I witnessed some relationships between women and men soldiers that become problematic. As a commander, I was not trained to deal with these issues. I found myself in uncharted waters overseeing awkward situations. Navigating these situations took time and energy beyond my normal command responsibility. Most military leaders tried to set a mature example. I referred to all soldiers by their rank and refrained from using gender terminology. More recently, I believe there has been considerable attention to these situations combined with sensitivity training. Yet, there remain unresolved challenges.

As the months flew by (no pun intended), we had many family visitors from the mainland. Jackie became a full-time tour guide and developed a standard itinerary. The Arizona Memorial in Pearl Harbor is a must-see for any visitor to Hawaii. While attending the University of Hawaii in downtown Honolulu, my son Dan became an expert on the best happy hour deals, which rounded out our tour agenda. Boating along the Waikiki coastline and beach trips to the north shore beaches were other visitor highlights. Watching world-champion surfers ride monster waves on the Banzai Pipeline was a singularly unique experience.

## My Next Bugle Call

As time drew toward completion of this impressive command experience, Brigadier General Brown called to congratulate me on being selected to attend the Army War College at Carlisle Barracks, PA. This was another triumphant bugle call. This selection paved the way for subsequent promotion to colonel and the possibility to command at an even higher level. The successive steps of being selected to attend the Command and General Staff College, promotion to lieutenant colonel, battalion command assignment, Regular Army appointment and selection for the Army War College were not something I ever anticipated or expected. My military career had taken an amazing about-face turn around in five short years. Professionally, I was on a roll.

After my command tour ended in April 1980, I was sent to Washington, D.C., as a member of a warrant officer promotions selection board. My report date to Carlisle was in June of 1980. Once again, Jackie managed our moving arrangements.

After we vacated our family living quarters at Schofield Barracks, we were authorized another brief stay at the Hale Koa Hotel before flying to the mainland. The boys appreciated one more opportunity with Waikiki surfing. Before we headed to the airport Dan, Dave, and Mike left individual claw marks in the sand. Our time in paradise had ended, but our memories have lasted a lifetime. Aloha.

# Chapter 15

# A Future in the Army

When the class of 1974 reported to the Army Command and General Staff College at Fort Leavenworth, Kas., we received a warm welcome. My 1980 arrival at the Army War College (AWC) in Carlisle Barracks was more like a VIP reception. The class was much smaller, with only 229 students. Of these, 160 were Army officers. The remainder was a mix of representatives from other services, reserves, civilians and 15 international students. The average age was 42.5. Everyone felt as if they had been hand-picked.

The class was divided into 16 numbered seminar study groups, each with 14 students. (If you're keeping track, this is a different total than above. A few seminars had an extra student.) I was assigned to seminar fourteen which included 10 Army officers and four other students. Our international representative was an officer from South Korea.

We were assigned on-base family housing. Our neighbors were friends from Hawaii, Terry and Kay Henry. The college included seven class members from my time in Hawaii. We had our own little alumni club.

Students and faculty commonly referred to our Carlisle experience as the "best year of your life." Even after living in Hawaii, the motto was appropriate. The biggest difference was the responsibility and stress of serving in command versus being a student. The Carlisle year was a singular opportunity to relax, take a good look at myself, and set future goals.

## An Amazing Education

The name Army War College (AWC) is misleading because we didn't study war. This happened at the Army Command and General Staff College. Carlisle College was formed in 1901 by Elihu Root. His original purpose statement was: "Not to promote war, but to promote peace, by intelligent and adequate preparation to repel aggression, this installation was founded." This principal concept has remained unchanged.

Carlisle Barracks has a unique history. An Army garrison was established there in 1757. In 1879, it became the Indian Industrial School and continued in this role for 39 years. As a federally-funded, off-reservation boarding school, over 10,000 Native American children from 140 tribes called Carlisle home. It became the model for 26 off-reservation schools in 15 different states and territories. A well-known Carlisle student was legendary athlete and Olympic champion, Jim Thorpe. The Carlisle area is rich in colonial and Civil War history, including nearby Gettysburg. The

Army War College was moved to Carlisle in 1951. Today, Carlisle is the oldest U.S. Army post.

I found Carlisle a brilliant backdrop for a year of professional development. Located in a historical setting, there was a combination of cultural and educational ingredients that combined for a remarkable experience. I am not aware of any comparable executive-level management training programs in the corporate world. In terms of professional education progression, it is the military equivalent of a PhD. Five such colleges are operated by the Department of Defense. Collectively, they are referred to as senior service colleges.

My timing for attending this schooling was pivotal, as 1980-1981 was a critical time for the U.S. But aren't they all? Our economy was in the tank, much worse than the 2023 economy. Double-digit inflation and interest rates strangled the consumer at every turn. Eventually, home mortgage rates reached 16 percent. The federal government seemed paralyzed to take any meaningful action. Iran held 52 Americans hostage for nearly a year after a raid on our embassy. The Soviet Union's massive military forces buildup posed a serious threat to our national security. For many in the U.S., this combination created a feeling of hopelessness and despair. Ronald Regan ran against President Carter in 1980. These factors contributed to very spirited and sometimes heated discussions in the confines of our non-attribution seminar environment.

Our academic year was divided into trimesters. The first section included a comprehensive examination and analysis of domestic issues, policies, and challenges. Each academic day began with a guest lecture by experts in their respective fields. Most days, we wore business attire rather than uniforms. The guest speaker's experienced views were often an eye opener, especially for students like me, who spent most of their careers in Army troop units. I was quite unaware of the world at the highest levels of government and business. Most of what I heard was extremely interesting.

After each guest lecture, students convened in seminar rooms for follow-up discussion. A member of each seminar was invited to attend a small question-and-answer session with the guest speaker. This person returned to our group and shared the question-and-answer highlights. Every day, we were afforded the opportunity to dig into the crux of critical issues and dissect the contents of the guest speaker's presentation. Our goal was to identify possible courses of action and create solutions to current and prospective problems.

For me, the discussions about the economy were especially interesting and important. Guest economists were all over the board with their opinions. Yet, there was one common thread: the number one priority for recovery was to restore confidence among the public at large. The fear of the current and an unknown future crisis held back the financial forces for recovery. Everything from consumer spending to corporate investment was on hold due to uncertainty. The unanimous solution to change this was national leadership and policies that would breed a new sense of confidence. This seemed so fundamentally simple compared to everything else the economists said.

It turned out they were right. During the 1980 election campaign, Ronald Regan portrayed a believable level of confidence that led to victory. In the following years, the nation made a remarkable recovery. The hostages were released in Iran. The economy gradually turned around. The U.S. began a military modernization and build-up that leveled the playing field with the Soviet Union.

Yes, there are different opinions about how much credit President Reagan should receive. Some economists believe routine economic cycles have little to do with government policies. With no horse in this race, I am merely relating what I saw and experienced. From the 1940's to the 1960's to the 1980's, national policies affected the circumstances I witnessed during each period.

At the onset of the school year, we were guided through a comprehensive process of self-assessment. This included a full physical evaluation and a diagnostic forecast of life expectancy. My prediction said I would live to 77. I could add another seven years to my life expectancy if I quit smoking. At the time, I was 42. It took another four years for me to accept this advice. At the time of this writing, I'm 86. I think it worked. Other diagnostic instruments provided an analysis of personality and leadership style. Most results were not particularly surprising but raised my awareness of my various strengths and weaknesses. This process helped me form a realistic understanding of who I was and where I was in life. This evaluation helped me believe what I thought I was capable of handling during the next ten years.

Statistically, we were told that one in four or five of us would be promoted to general officer. We were also told that most of us were similarly qualified and would serve as Colonels in positions of the highest responsibility throughout the Army for most of the next ten years. In these positions, we would have the most day-to-day direct influence in running the Army. It's likely that most of us initially assumed we would be in that top 20 percent, even though the odds were against us. I don't have a final total, but I think the math held up pretty well. Three class members became four-star generals. One of them, Richard Meyers, became Chairperson of the Joint Chiefs of Staff. As for myself, this was the first time in my career that these lofty goals were even discussed. Now, I was in the big league where anything was possible.

Our initial seminar group remained together until the holiday break. After the holidays, we were reassigned to another seminar group with a completely different student group. Initially, I was disappointed because of the class intramural sports program. During the initial term, there was a softball competition. In the second term, it was basketball. My first group had four of the tallest students. Fortunately, my second group had the best and tallest basketball player.

The second term focused on international issues and policies. We took a hard look at where we stood in the world and the major elements of national power. We examined military, political, and economic components in detail. These studies revealed that economic power, rather than military power, is most important.

As we were wringing our hands over the Soviet Union's military threat in 1980, global experts pointed to a different long-term scenario. Advisors shared with us the prediction that within 50 years, China would be the greatest threat to our national security as an economic superpower. I also recall a prediction that eventually Russia could become a U.S. ally to combat Chinese global aggression. Fifty years later, we are in the predicted window of China's economic growth. While there is no apparent likelihood of any partnership with Russia given its invasion of Ukraine, China's recent military actions tend to confirm predictions of aggression. I have not returned to Carlisle since 1981. It would be fascinating to sit there for one day and hear the current classified intelligence briefings and assessments.

In the final phase of our academic year, we rejoined our original seminar group and worked on developing strategies for U.S. global policy and national security. In our previous studies, we saw how much of our nation's policy development was short-term. Most policies do not extend beyond the current administration and budget cycles. Our major adversaries, Russia and China, have exceedingly long-reaching policy goals. We appreciated the important value of strong regional partnerships that share common national interests and focus on economic, technological, and military capabilities.

The North Atlantic Treaty Organization (NATO) alliance's long-term success is an important model to apply to other regions of the world. Since 1981, the emergence of similar regional partnerships has provided long-term policy formulation that endures changes within the U.S. government leadership. The continued growth, development, and strength of these commitments are essential to our future national security.

## Time for a Master's Degree

An added benefit of attending the AWC was the option to participate in a cooperative master's degree program with nearby Shippensburg State University. This was an excellent opportunity to acquire a graduate degree, which two-thirds of my classmates already had. In the program design, Shippensburg awarded credits from some of the Carlisle courses. The remainder of the classes required in-resident study at their main campus for two summer semesters and twice weekly classes during the Carlisle academic year. It added a considerable amount of study, but a far more manageable schedule than what I endured at Fort Rucker, Ala. in 1967.

I chose communications as my major in the university's journalism department, based on my curiosity about mass media communication's impact on public opinion. When I left Vietnam in 1970, I had a low opinion of how the war was reported by national mass media, especially television. The reporting failed to separate the war from the warrior. It focused on the futility of our efforts rather than what we accomplished. Soldiers felt we performed our jobs very well. Everything in the media reported our efforts as failures. The creation of stereotyped soldiers' images unfairly distorted the truth. This continued after the war with negative portrayals of veterans and sensationalized reporting of how we were allegedly treated. While there were documented incidents of returning soldiers being mistreated, I can honestly report that among the thousands of veterans I personally encountered, I have yet to find one who was spit upon. All these thoughts ate away at me. I was anxious to get to the bottom of why these things happened.

Another interest was interpersonal communications. A favorite and frequently heard saying was, "The boss never listens to me." I was fascinated with the power of persuasion. Ultimately, I wanted to influence decision-making successfully. This was quite different from the mass media issue and a separate path of research and investigation.

Both areas led to surprising conclusions. Some of my previous opinions were altered. Elective Carlisle courses, apart from the Shippensburg course requirements, were also enlightening. These focused on the specific military interaction with reporting entities. One example was an elective called, "The Commander and the Media." The class consisted of a progressive mock interview series that was filmed and critiqued by fellow students. We studied tell-tale facial movements and body language mannerisms in reaction to questions. The final episode was a mock press conference with an amazingly effective group of staff volunteers acting as a press corp. We were required to make an initial statement about a serious incident on the base. Upon entering the room, mayhem erupted with shouted questions, flashing cameras and thrust microphones. This was all captured on video.

Based on my education, what were my conclusions? Regarding mass communications, I developed the important understanding that major networks are in an incredibly competitive entertainment industry. Their success is dependent upon viewer ratings and sponsor advertising. It's a business with no responsibility to report anything that doesn't sell. Their primary effect on viewer opinion is to set the daily agenda of what the public hears and talks about.

Our actual opinions are formed more by our discussions with others. We are not nearly as brainwashed by the media as we are led to believe. I think the situation has steadily declined over time. Currently, there is more sensationalizing and bias than ever. It's become more overt with a lot of openly political reporting. It's hard to find programming that reports news rather than opinions. Reporters are known to the industry as talent. Their physical appearance and personality are increasingly more important. In the 1960s, many people listened to Walter Cronkite for the evening news. Everyone believed what he said. There was nothing flashy about Walter. He seemed so genuine. He concluded each news broadcast saying, "And that's the way it was." The report was the way Walter said it was, but was it true? A close evaluation might reveal a bit of his own subtle bias.

The AWC experience included considerable social life and recreation. From welcome receptions at the college president's home, to Army, Navy, and Air Force Balls, we also enjoyed seminar group parties and more. There were class field trips to Washington D.C., New York City, and Gettysburg. In New York City, two Shippensburg classmates and I had a private meeting with the *Newsweek* leadership team, including the editor-in-chief. This led to a very enlightening discussion that enriched our knowledge of the major news outlet's inner workings. A good case was made for how print media typically contains more detail and objectivity than electronic media. Printed news also permits more opportunity to digest content than the rapid sound bites of television news. Still today, I prefer to read the news than listen to it.

Recreation was another important aspect of AWC life, especially intramural sports competition. Our class contained a very fit and athletic group of 40-year-olds. With our star basketball player Jerry Bell, my seminar team sailed to the final four championships. We seemed unbeatable. In the semi-final game, Jerry sprained his ankle early in the game and could not play. We gave it our best without him but lost by three points to the championship team. This scenario reminded me of my last college game in 1958 and our mantra said, “It ain’t over until it’s over!”

At the end of the year, All-Star teams competed in an annual inter-service event called the Jim Thorpe Sports Days. Held prior to graduation, the five senior service colleges assembled at Carlisle to compete in a variety of team and individual sports. Points were awarded for placing in each of the 12 events and determined the final standings and overall winner. In previous years, the Army dominated and won the overall championship. Some felt our home-field advantage influenced this. In 1981, the word was out that the Air War College was loaded to win. It came down to the last event: basketball. After playing a double-elimination tournament, Army and Air Force were tied for first place. One more game decided the winner. I was a member of the basketball squad and recalled our college president coming to our bench at half-time. His message was short and clear: if we did not win, we would not graduate! The game remained close until the end, but we managed to pull out the win. Whew! We could graduate!

The Carlisle year was important for our son Mike. He attended his high school junior year and made a lot of new friends. Mike blossomed on the basketball court and enjoyed it. I coached his youth sports team and encouraged him. Mike learned to play the guitar while we lived in Hawaii and followed up by taking lessons in Carlisle. Mike and a group of friends formed a rock band which became a big hit with the Carlisle teenage crowd. He practiced playing without coaxing. His band was dedicated to doing the same. Unfortunately, their self-discipline was not rewarded with a place to practice. Jackie and I decided their effort to become good at something was worth nurturing. We opened our home as a band headquarters.

The band practiced often. Jackie and I witnessed how surprisingly diligent they were to achieve perfection. AC/DC was red hot in those days. The boys practiced each song until they sounded exactly like the recordings. It was amazing. We realized these guys worked every bit as hard as our other sons did in football. They were doing it on their own without coaches!

From this experience, I learned the value of teenagers finding something positive to become good at and excel. Mike became a good musician. He developed a sense of responsibility that lasted him a lifetime. Every kid cannot become a top student or athlete. There are many other interests that can produce a positive outcome. Kids who lack a compelling interest may be more at risk than those making a lot of noise with guitars. Mike also met his first love this year. Like his brothers before him, he crashed hard when it was time to move and start all over. Contrary to popular belief, I discovered that boys land hard when these sad endings occur.

## Exploring My Next Assignment

During the months leading up to AWC graduation, my next assignment became an important priority. For the first time in our careers, AWC graduates did not have to accept any assignment. We were afforded choices. For some students, this provision led to extensive negotiations to find the right job. In my case, I didn't think my next assignment mattered too much because I was likely headed for promotion to colonel as an AWC graduate. Honestly, I didn't see myself in the top 20 percent and headed toward general officer rank. I had already progressed farther than I anticipated.

Two separate assignment officers contacted me, both I'd served with in the past. One suggested I accept a job in San Antonio, Tex., at the Fifth U.S. Army Headquarters. The other officer urged me to accept an assignment at the Pentagon. He explained the importance of this opportunity. I chose to reject his advice. In my heart, he was probably right. But this was a bugle call I didn't believe was calling me. Instead, I took the job in Texas.

This remarkable year ended with a graduation ball and commencement exercise on the historic campus grounds. With the sun shining brightly on a beautiful June day, my classmates and I assembled for the last time. I don't remember anything the main speaker said. I do remember looking at my classmates and feeling extremely fortunate to have spent a year with this incredibly elite group of outstanding people. The memories we shared, and our combined learning would last a lifetime. This experience was a true blessing.

As we shook hands and wished each other well, I was reluctant to bid farewell. My thoughts returned to something I learned in Germany. We never say, "Goodbye."

Instead, we said, "Auf Weidersehen." Until we meet again.

## Remember The Alamo

San Antonio isn't exactly Hawaii, but it is a wonderfully comfortable place to live. In 1981, I was told the only location with more retired generals than San Antonio was the Washington, D.C., area. Multiple Air Force bases and the Fort Sam Houston Army base attracted many retired service members. This area can grow on a person with the historic Alamo Fort and the unique River Walk area which draws tourists as it winds through the old downtown area. The 750-foot-high Tower of Americas provides spectacular views. San Antonio is a neat place to visit, and it attracts many conventions and conferences. During the three years we lived there, we became a magnet for visitors. Jackie renewed her tour guide role.

The Fifth U.S. Army headquarters was located in a historic quadrangle fort facility. Completely enclosed, it housed an assortment of wildlife that roamed the grassy area in the middle. There was limited vehicle parking for senior staff members and a large outside parking area for other employees and visitors. The clock tower in the quadrangle is another landmark. We were assigned to family housing in a historic building erected in 1901.

My assignment was quite different than anything I had experienced. Commanded by a three-star General, its mission focused on the training and readiness of reserve component organizations in a thirteen-state area that stretched from Texas to Ohio. Oversight included the National Guard Headquarters in each state and approximately 20 major U.S. Army Reserve commands. I don't recall the total number of troops, but it was many thousands. In 1981, the Fifth Army had considerable oversight responsibility among these organizations, especially the Army Reserve Headquarters. Budget management was a huge mission.

I was assigned to the office of the Deputy Chief of Staff for Training (DCST), which was headed by a colonel. As a lieutenant colonel, I was responsible for one department division. My situation caused me to reflect upon my early days as a National Guard soldier. I saw the biggest picture of what the reserve components included. At the time, there were three continental U.S. Army headquarters. The First Army was responsible for the eastern United States. The Fifth Army oversaw the central states, and the Sixth Army covered the west. Each headquarters had subordinate headquarters known as Army Readiness Regions commanded by two-star generals. The regions had subordinate units called readiness groups commanded by a colonel.

My job didn't seem to require a War College graduate. In fact, my duties lacked relevance. My boss had been there long enough to become fully disillusioned. Once, he told me that we could walk away from our office and return three months later, and no one would know we were gone. He was being facetious, but he did have a point.

Today, I understand the headquarters' mission is different and more important, but I lack current details. In March 1983, I was promoted to colonel and became head of the DCST office. At about the same time, the Army created an aviation branch and completely restructured how Army aviation was organized and managed. Prior to this, the federal Defense Officer Personal Management Act legislation was enacted. This Act completely changed how reserve and regular Army officers were managed. It eliminated the indefinite active-duty category I had struggled in for so many years. These two changes combined to move the system positively forward.

In addition, a new Active Guard/Reserve category was created. This enabled reserve component officers to serve in active-duty status for up to 30 years but in assignments separate from those managed by the active Army. This may all sound confusing. Suffice to say, the Army has a far better system today than when I began in 1961.

Through the DCST, I had considerable direct contact with senior Reserve component leaders throughout the Fifth Army area. These personal relationships granted me a much greater appreciation for their dedication to military service while maintaining significant private sector positions of responsibility. I enjoyed collaborating with them and building new friendships. Some connections came back to roost during my final Army assignment several years later.

As my understanding of the full complexity of the Reserve component world grew, my appreciation for their service grew as well. They had a tough job. I took my duties seriously and tried to be part of their solutions. At least some of those efforts may have been useful.

In 1984, a new brigade command selection list was released. I was listed as an alternate designee. In somewhat of a replay of my battalion command selection, this time I knew I was selected. I was pleased. I also knew an unforeseen vacancy would have to occur before my activation. I saw the potential brigade command as a fitting climax to my long journey as a soldier; one that would be an honor and a privilege. I could only wait and see if it would happen.

Shortly after receiving this news, I received a call from Washington D.C. The assignment officer did not inform me of a command opening. Instead, I had been nominated to serve as the Director of Advertising for the U.S. Army Recruiting Command (USAREC).

My response was, “WHAT?”

In a computer search with the parameters being that of a colonel with a master’s degree in communications who was an aviator with combat experience, my name was the only one that came up. Go figure.

This assignment would mean moving to Fort Sheridan, Ill., where the USAREC headquarters was located. I qualified for on-base senior officer living quarters, some of the nicest in the Army. My assignment included one year of industry training with a major downtown Chicago advertising agency. Upon training completion, I would take over the USAREC account and manage it for the next two to three years. Equipped with managing one of the largest advertising accounts in America, I could then retire from the Army and, in his words, "name my own price" in the corporate advertising world.

At this time, the "Be All You Can Be" Army recruiting commercials were on television every night. Everyone knew the music. USAREC was preparing to find a new theme. I think the intention was to highlight the excitement of helicopters. When I told my close associates about this development, there was unanimous agreement. This was a once-in-a-lifetime opportunity, an offer no one could refuse.

The head of USAREC, a two-star general, requested an interview. He was a field artillery officer I knew of, but not personally. During the interview, I explained to the general that I was on standby for a brigade command assignment, something I wanted to do very much. He had carefully reviewed my entire service record and knew a lot about me. In an honest appraisal of my circumstances, he said if I received a brigade command job, there was no guarantee I would be promoted to brigadier general. My age and lack of Pentagon experience worked against me.

If I took the advertising job, future advancement was a sure thing. This sounded more and more like a no-brainer decision. I appreciated his candor and thanked him. Yet, I knew my calling had always been to be a soldier. I had never been motivated toward lucrative rewards in the corporate world. After further consideration, I respectfully declined the advertising job and took my chances on getting the job I really wanted.

This was a huge bugle call I chose to ignore and accept whatever consequences followed. Did I make the wrong decision? Perhaps. I also believe that once a major decision is made, it isn't helpful to look back.

Not long after this unusual experience, I received a call that informed me that I was activated from the alternate list to a command assignment. The Third Infantry Division in Germany had received approval to activate a newly created Division Aviation Brigade. All the primary selectees were programmed for other assignments. I was up next.

An Army leader in becoming fully modernized, the Third Infantry Division wanted to be the first division to activate an aviation brigade. In another unique situation, I was honored to be chosen. The Division had already begun to organize provisionally. I was immediately contacted by the Division and the aviation team. Activation was scheduled for March 1985. This was a two-year assignment.

The new assignment meant another round of pre-command training. A new requirement for Aviation Brigade Commanders required that we be qualified as instructor pilots. I would actually be the first person to attend the course. This meant a longer temporary duty stay at Fort Rucker.

As we began moving preparations to Germany, family consensus decided that Texas would become our long-term home. Dan and Dave were both married and enjoyed living in the area. Jackie and I also liked it. We bought a home where Dan and their family would live until we returned to the States.

Jackie and I tried to imagine what it would be like to return to Germany after 20 years. We knew it had to be more enjoyable than my earlier assignment. We looked forward to this new adventure. This time, I would have the responsibility and resources to ensure every aviator received their required training and maintained their qualifications. This was a good feeling. I was anxious to begin.

# Chapter 16

# Rock of the Marne

The Third Infantry Division has an important combat action history that began during World War 1. While defending against superior German Army forces along the Marne River in France, they fought fiercely and held the line for the Allies. This battle earned the division the nickname of Rock of the Marne. The Third Infantry Division fought in World War II, covering ground from North Africa to Sicily, Italy, Germany, and Austria. Soldiers saw more than five hundred days of continuous action. Again, they were called to serve during the Korean War from 1950 to 1953. During the Vietnam War, the division was stationed in Germany. Since then, the division has remained forward deployed along the European freedom frontier.

My 1985 arrival in Germany went much more smoothly than in 1964. This time, all the right people knew our schedule and were expecting us. We were met at the Frankfurt airport by a staff member of the Aviation Brigade and taken to Leighton Barracks in Wurzburg, the Third Infantry Division's headquarters. Temporary housing awaited us at the base guest house. We also had access to dedicated transportation to get us around. Initially, it was just Jackie and I starting this brand-new adventure, as Mike arrived later.

Soon after arrival, I met with the Division Commander Major General Howard Crowell. He expressed a warm welcome and showed genuine enthusiasm for the new Aviation Brigade's potential. In Vietnam, Crowell served as an Infantry Battalion Commander in the 101st Division, which meant he had significant personal experience with helicopter capabilities. The Division Chief of Staff, Col. Roger Bean, was also a 101st Vietnam veteran and fellow Field Artillery Aviator. The Division Operations Officer, Lt. Col. Rik Shinseki, was another Vietnam combat veteran. Eventually, Shinseki became a four-star General and served as the Army Chief of Staff. It was reassuring to have their deep experience which provided me with significant support during the challenges I faced.

New battalion and brigade commanders were required to attend a one-week orientation session at the Grafenwoehr training center. This provided a big-picture view of the U.S. force's mission in Germany, along with an understanding of how our command assignments fit in. In 1985, the Cold War was still very real. The Soviet Union posed a serious threat. The U.S. Army was in the midst of a significant modernization of its warfighting capabilities. The Third Infantry Division was at the forefront of implementing this new equipment and technology.

At the orientation session I attended, two of my future helicopter battalion commanders were present. We became acquainted. At one of the briefings, Major General Stotser, U.S. Army Europe Headquarters Operations Officer, spoke. He was scheduled to take command of the division. It was interesting to hear his perspective about the challenges we would face.

Upon completion of the orientation, we returned to Leighton Barracks and were immediately assigned family housing. It was a recently renovated single-family house located on a street with other senior officers, including the commanding general. It backed up to a large green space that was originally a German military airfield. The Leighton Barracks complex included a large commissary grocery store, post exchange facilities, a gas station and an auto repair shop. I don't recall the total population living and working here, but with multiple organizations, it totaled several thousand. The compound included schools for military-dependent children. The Wurzburg Army Community also had its own full-service hospital and dental clinic.

## Local German Culture

The city of Wurzburg is located within a region known as Franconia which is situated along the northern border of the Bavaria state. This entire area has a deep and ancient history. Wurzburg is geographically centrally located with subordinate division elements in outlying areas. A combat brigade was stationed in each of three other cities: Schweinfurt to the north, Aschaffenburg to the west, and Kitzingen to the east. The aviation brigade was located at an airfield adjacent to the small city of Giebelstadt, slightly south of Wurzburg. The unique Franconian culture includes the local vineyards that produced Frankenwein. It is the only German wine that is decanted into the distinctive "Bocksbeutel," a small, rounded bottle. This wine has an unusual taste that takes a little getting used to.

The Giebelstadt Airfield also has a special history. Established in 1936 for the German Air Force, it was kept secret throughout World War II. It was cleverly designed and camouflaged to prevent air detection. The airfield and the city of Giebelstadt were deleted from all World War II maps. Even when it was eventually located, it was never bombed. Near the end of the war, some of the first German jet-powered aircraft were based there. After the war, the base was never turned over to the U.S. Instead, it was retained as a NATO asset. While it was maintained by the German government, the U.S. Army used it. Briefly, the U.S. held German prisoners of war on the base. By the time I arrived, one of those prisoners had become the mayor of Burgermeister of Giebelstadt.

These geographic and historic details of the Wurzburg and Giebelstadt area helped establish the environment where I served as a U.S. Army brigade commander for two years. A careful understanding of German culture and customs became essential items in my commander's toolbox. Because the air base was so close to the small local community, we were joined together at the hip and shared all aspects of daily life. The local civilian population was approximately 3,500. With families, the Army population was similar.

This setting created some new challenges for me. It also afforded a remarkable opportunity to establish unique working relationships and wonderful friendships in a foreign country. While there were ample soldier barracks on-base, all of the families lived off-base. Some lived in military housing located in Giebelstadt. Others lived in civilian apartments scattered throughout the surrounding villages. This resulted in a widely spread integration of culture and mutual understanding.

As guests in a host country, maintaining smooth community relations was an important command responsibility. German and American relations included the operative terms of *partnershaft* (partnership), and *zusammenarbeit* (working together). While these are not particularly unique concepts, try to picture the mixed culture environment. It was essential for me to establish and maintain a close working relationship with Mayor Werner Mantel. One major problem: he spoke no English. Not to worry, Ray! A German employee on the base was assigned additional duties as my translator. Initially, I used him until I discovered that he and the mayor were on different political pages. The mayor did not want him involved in our discussions. This was especially sensitive during public gatherings when I tried to express my appreciation for the mayor's assistance. So, I dusted off my high school and college German studies in a hurry. The mayor appreciated my efforts and encouraged me to do the best I could. We worked it out.

Mayor Mantel became a super partner. Jackie and I befriended him and his wife, Rosie. We frequently attended events together. Another important friendship evolved with the town doctor and his wife. Dr. Otmar Pfeiffer was highly respected by everyone in the community and an influential opinion leader. In time, Dr. Pfeiffer and his wife, Ilse, became our best friends in the community. We shared many good times. This became a long-term relationship and I still try to stay connected with them.

The Pfeiffer's taught us a lot about the German healthcare system. His responsibility area included six villages in the Giebelstadt area. He received an annual budget from the government to provide primary care to everyone and did not charge for these services. Specialty care was provided at the Wurzburg civilian hospital. Outside of Germany, this country-wide system is sometimes referred to as free health care. Everyone and their employers pay into it, based on individual income. It is affordable, but not free. This was a no-frills system that avoided unnecessary testing and treatments. Dr. Pfieffer and his wife, a nurse, operated the Giebelstadt clinic which included lab testing. He made house calls to housebound patients. At the time, most of the widowed elderly resided with family members. Dr. Pfeiffer was on-call, 24 hours a day, 7-days-a-week, which was shared with another area doctor.

Another important community member was Maya Pohl, the director of the Red Cross office. Along with managing blood drive events and other routine services, she was a great friend to those in need, especially younger American soldiers and spouses. She ran her own system for acquiring donated household and personal items. One of Mayor Mantel's favorite sayings suggested if someone couldn't find something anywhere else, "Go see Maya Pohl."

Quickly, we realized it would be much better to live in Giebelstadt than the military housing at Leighton Barracks. While only twelve miles apart, the symbolism of living among my community partners as well as the soldiers and families under my care was an important message to everyone. Dr. and Mrs. Pfeiffer helped us find a suitable place to live in Giebelstadt. They found a lovely duplex home owned by a university professor. He planned to be away for an extended sabbatical and was interested in renting to the right party. The Pfeiffer's vouched for us. We worked out the details with the Army and made the move. We immediately felt the results of being embedded within the community. I was also more responsive to the headquarters.

Baron Stephan Zobel was another important personality in Giebelstadt. He was the last family descendant that originally owned the land where the city stands. He and his wife lived in a castle in the city center that dates back to 800 A.D. It was and remains in a state of long-term restoration. The Baron had a flamboyant personality but was most gracious and hospitable. He was also very patriotic and a major in the German Army Reserves. He owned an older classic Porsche sports car, which I tried to buy. Unfortunately, the car was near and dear to his wife who refused to part with it. On one occasion, she gave me a demonstration ride. When she laid a strip coming out of the castle onto the main road through town, she did not leave the best impression on the townies. This scene did not match the lofty expectations they had for the image of a colonel.

## Settling into My Role

In March 1985, the Brigade's activation ceremony was very impressive. Displayed helicopters spanned the skyline and provided a backdrop behind the assembled soldiers standing in formation. The brigade cavalry squadron served as the ceremony's color guard. They wore old cavalry uniforms, complete with black Stetson hats. Boy, did they look great! All of this made for a proud presentation of what the brigade represented.

Major Jim Mowery organized the ceremony and did a remarkable job of pulling it all together. The division commander chose him because he had experience from a previous assignment. On activation day, he became the commander of one of our attack helicopter battalions. His hard work paved the way for a smooth transition for me because so much of the heavy lifting was completed. Jim was an outstanding battalion commander, which was reflected in his selection to attend the Army War College the following year.

Most of the brigade was stationed at Giebelstadt. The brigade included two attack helicopter battalions equipped with anti-tank Cobra helicopters, one UH-60 Blackhawk company, a general support company with UH-1H Huey helicopters and OH-58 Kiowa aircraft, and a support maintenance company. The cavalry squadron was based at Schweinfurt and included two air troops companies that were equipped with Cobras and OH-58 Kiowa helicopters, along with two ground troops equipped with Bradley fighting vehicles. The squadron had a 24-hours-a-day, 7-days-a-week mission to patrol a sector of the east-west German border. We used Coburg as a forward operating base, which was located north of Schweinfurt and adjacent to the border. The cavalry squadron pilots were the only ones in the brigade who were trained and qualified to fly along the border and avoid overflight.

The Giebelstadt base was self-contained and designated a sub-community of the greater Wurzburg community. It had a commissary, a pharmacy and a snack bar in addition to excellent recreation facilities and a health care clinic. There was even a four-lane bowling center. The Giebel People Inn, a community club facility, was a focal point for social life. Altogether, Gieb was a cozy setup for our soldiers to live, work, and play. A current Facebook page for Gieb alumni features frequent soldier postings expressing fond memories of days gone by. The U.S. Army ceased operating there in 2006. Today, the vacant buildings remain intact but are surrounded by a fence. Civilian aircraft use the runway.

Throughout my career as an Army officer, I've always seemed to be in a place where major changes and innovations took place. This was repeated during my brigade command assignment. In 1985, the Aviation Branch was quite new. Major changes in organization structure and employment doctrine unfolded. New thinking exploded during my Command and General Staff College experience 10 years earlier. These ideas still played out. Army Aviation leaders believed the new combat aviation brigades shared equal status with the other brigades in each division. It meant they should become known as "maneuver" brigades rather than "support" organizations. In the past, aviation battalions in each combat division were considered support elements. These distinctions may seem trivial, but they produced considerable resistance among some non-aviation leaders.

The debate centered around the belief that an aviation brigade could not occupy and hold ground like traditional infantry and armor brigades. This was a valid point in terms of traditional paradigms, but not necessarily absolute on the modern battlefield. Initially, I convinced the division commander to designate aviation as the 4th Brigade of the Third Infantry Division. When other divisions activated the aviation brigade, they did the same. Others did not. To my knowledge, the Army never took a position on this. I recall top Army leadership directed references to the aviation brigade as a maneuver asset should be omitted in all the Army field manuals.

Another pushback involved the cavalry squadron. Cavalry leaders held a major contention that a cavalry squadron had no business being in an aviation brigade. Traditionally, the cavalry squadron was a separate unit under the direct control of Division Headquarters.

In a nutshell, these parochial attitudes put me in the position of having two years as the first commander to demonstrate the value of these new concepts. This would be no easy task.

The first confrontation wound up in the division commander's office, who was asked to resolve whether an aviation brigade was a maneuver asset. The division commander reminded everyone present that the technical definition of maneuver included both fire and maneuver. Since the aviation brigade had both, it could legitimately qualify as a maneuver brigade.

In December 1986 during a major field exercise called Winter Warrior, my brigade was assigned the mission of division reserve with standby counterattack responsibility. To accomplish this, two ground battalions were attached to my brigade. The arrangement proved advantageous. Even the tank battalions thought things went smoothly. Honestly, I don't know where this stands today in terms of current employment doctrine and policy.

During my command period, significant technological improvements took place. Night vision goggle use was implemented throughout the Army which required significant training and practice. We also prepared for the fielding of the AH-64 Apache helicopter to replace the Cobra. Construction of new hanger facilities was underway. Giebelstadt was also home to an air defense missile unit. The Patriot air defense system fielding was on the way. I received an orientation flight in the Apache and could readily see its vastly improved capabilities over the Cobra. During the flight, I felt the realization of serving as an Army aviator from the 1960s era Wright Brothers technology in the L-19 Bird Dog to the 1980's era of Star Wars technology in the Apache. It was a startling experience and contrast.

## Drinking in the Local Culture

Living in Germany for two years allowed us to experience German culture. A few stand out. Most activities are family-oriented. Even at community festive events, people of all ages have fun together. The liveliest time of the year is Fasching, an equivalent to Mardis Gras in the U.S. These pre-Lenten rituals are similar, but in Germany they are celebrated equally throughout the entire country. Every community, large or small, has parades and Fasching Balls. If someone walks with a paper cup in hand during the parades, someone is apt to pour wine into it.

In German culture, crime prevention is considered everyone's responsibility. Crime, in any form, is considered unacceptable. In America today, crime feels like it is growing out of control. While the public is urged to report suspicious activity to police authorities, unfortunately, this does not always happen. In Germany, citizens readily do this without prompting. They notify local authorities of any suspicious activity. Even minor infractions, like honking a vehicle horn unnecessarily, are reported. This may seem extreme, but it was a successful deterrent.

Law enforcement jurisdiction for Giebelstadt was headquartered in the nearby city of Ochsenfurt. Police Chief Herr Frank was another important person for me to know. As he quietly made his way around his beat, he regularly received an earful of information about unusual activity. When I talked with him, I was always amazed at how much he knew about the off-duty conduct of my soldiers. He shared any potential concerns for us to look into. This was another partnership that helped avoid problems.

Noise abatement was another important German public concern. In many small towns, shops closed on Saturday afternoon and remained closed until Monday. Sunday is strictly a quiet day of rest, worship, and family activities. It was common to see families walking together. In eating establishments, voices were restrained and seldom became loud or boisterous. It was considered bad manners to make noise in public places, except during Fasching. Good manners were always expected. Sometimes, I am reminded of this etiquette when eating in the U.S and I hear a group talking and laughing in a loud or obnoxious manner. Customary greetings always recognize women first with handshakes. These customary behavior examples are indicative of a general state of order throughout German society.

## Accidents Happen

Our brigade flew thousands of accident-free hours and was a leader amongst aviation units in established measurements for safety and readiness. Nonetheless, we had two major training accidents near the end of my assignment. Both were a result of pilot error. Aviation mishap prevention is a full-time challenge that requires diligent concentration. Training for combat missions has increased risks compared to flying straight and level at safe altitudes. Much of the tactical training performed at the Grafenwoehr training area included live firing from the Cobras. The weather was a frequent challenge, just as it was 20 years earlier. I did all of my flying in the UH-1H Huey which was equipped to fly during instrument weather conditions. There is an advantage in being able to land anywhere and fly at slower airspeeds when necessary.

Each of the four brigade headquarters took turns representing the Third Infantry Division headquarters command post at Grafenwoehr. Duties included hosting official visitors who frequently came to observe training and the performance of the M-1 Abrams tank. This responsibility allowed me the opportunity to receive firsthand familiarization with the M-1 tank, including firing the main gun system while moving and firing at a moving target. I was amazed at its deadly accuracy and quickly realized how superior it was to any other tank at the time. One visitor was the Chief of Staff of the Israeli Defense Forces. When departing, he presented me with a commemorative coin. It remains a special item in my collection of military memorabilia.

During my frequent absences from our Brigade Headquarters at Giebelstadt, the Brigade Executive Officer, Lt. Col. Jim Neidig, was in charge. He did a wonderful job of keeping everything on track and coordinating with battalion commanders. When I decided to provisionally form a fourth battalion by combining separate companies within the brigade, Neidig was the perfect choice to command this battalion. He and his wife Cindy became particularly good friends.

Other key brigade players were Major Jim Simmons and the Brigade Command Sergeant Major Othell Terrell. Simmons was the brain behind the creative thinking for the brigade assets tactical employment in battlefield scenarios. Subsequently, he rose to General Officer rank. Terrell, a fine soldier, provided the glue of effective communications among the ranks of senior non-commissioned officers. The chain of communications he managed was key to our effectiveness. The chain of command is most effective at disseminating orders and directions downward and also inhibits the upward flow of communications that can be stopped at any level. The senior non-commissioned officer chain can keep the command chain informed of misunderstandings or disconnects at any level. Terrell and his wife Pat were a notable example of an Army couple who worked as a team to support everyone.

Advisory councils can also enhance effective communication within an organization. Dedicated forums that allow for open discussion among diverse groups of enlisted soldiers and officers provide auxiliary communication avenues. When effectively managed, they can strengthen the health of any organization. Building mutual trust and respect at all levels within the organization is a huge multiplier of success.

Brigade Chaplain Captain Paul Howe and his wife Lois were another all-American team who touched the hearts of everyone in the community. A wonderful pastor, Howe was a reliable source of spiritual strength during our most difficult moments. I have never forgotten his admonition that a benediction does not signify an ending, but the promise of a new beginning. His words rang true to my belief that the bugle playing of *Reveille* and *Taps* meant the beginning of each day with an ending that promised another new day. Lois, a nurse, worked at our health care clinic and was fondly known to everyone. While I have mentioned only a few people I relied upon every day, there were many, many, others.

The brigade commander role is vastly different from my descriptions of company and battalion-level command. It is essential for a brigade commander to fully delegate responsibility to his staff and subordinate commanders as well as provide necessary support. Battalion commanders are responsible for the integration of planning and necessary functions to accomplish each mission. They must conduct daily missions while also seeing to the health and welfare of the soldiers and families under their care. Success at each command level produces a winning team. The necessary 24-hours-a-day, 7-days-a-week command responsibility required to accomplish these objectives is unique to the military culture.

Corporate world success stories where employees take specific steps to care for their employees and family members warms my heart. This kind of culture makes sense.

As my two years of Brigade Command approached conclusion, the question of my next assignment came into view. A commonly held belief among the Army corporate body was that the assignment immediately after brigade command was a most crucial step toward General Officer promotion. Initially, I had an interest in remaining in Germany and being assigned as a chief of staff at a higher headquarters. Instead, I was informed the Chief of the Army Aviation Branch wanted me to be assigned to the U.S. Forces Command Headquarters (FORSCOM) in Atlanta, Ga. This command included control and staff oversight of all combat aviation assets stationed in the U.S. By 1987, I had become one of the most senior-ranking and experienced aviation colonels in the U.S. Army. The FORSCOM job made sense. I signed my name on the dotted line.

I had a far different feeling leaving Germany in 1987 than I did in 1966. After my earlier assignment, I couldn't wait to leave Germany. Now, I hated to leave. We left so many wonderful friends and experiences behind. We enjoyed German culture to the fullest.

I also walked away from the stark realities and dangers of the German border threats. Meanwhile, our German friends said they could hear the winds of change blowing from the east. Our helicopters reported sounds of mines being detonated along the border fencing barriers that extended the Berlin Wall. It still seemed unimaginable that Germany could be reunified in the near future. I entered active-duty in 1961 when the Berlin Wall went up. Was it possible I would see it come down before my service ended? Could the long Cold War end without firing a shot? All these thoughts churned in my mind.

At my change of command ceremony in March 1987, "Auf Wiedersehen" was the common expression shared. We had such sadness in our hearts as we gave our final hugs to our dear friends, the Pfeiffer's. We knew we would always be friends and miss each other very much. The memories of our two years together lasted a lifetime.

In June 1987, not long after we left Germany, President Reagan delivered a speech to the German people that included the famous statement, "Mr. Gorbachev, tear down this wall!" I still wondered if this could actually happen.

Two years later, the wall came down. The Cold War ended. Soon, there would be another major shift in U.S. military operations.

# Chapter 17

# A Dangerous World

When I returned to the U.S. from Germany in March of 1987, it prompted a wave of new personal thought and reflection. As much as I enjoyed living in Germany, the experience may have caused me to temporarily lose track of the life I cherished as a U.S. citizen. Stepping back into American life was like flipping on a light switch. I rediscovered how fortunate I am to live in our great country. This was a transition between two different worlds. Each has its unique characteristics and advantages. At the end of the day, I strongly believe there's no place like home. The special American spirit has no equal with a foundational value of freedom and equality that I have not seen elsewhere.

During my two years in Germany, I was so focused on the military threat we faced there. Upon re-entry into the U.S., I realized that I had also lost sight of what was happening in the rest of the world. Many possible scenarios I was exposed to at the Army War College had drifted from my awareness. My assignment at the U.S. Forces Command Headquarters (FORSCOM) brought current global realities sharply into view. I became immersed in issues I didn't realize existed and gained a full understanding of the worldwide dangers the U.S. faced.

In 1987, the FORSCOM headquarters was on a small installation in Atlanta, Ga., named Fort McPherson. Four-star General Joseph Palastra commanded the headquarters and included several other staff general officers. As the staff aviation officer, I reported to the deputy chief of staff for operations (DCSOPS), a two-star General. The headquarters total size and scope was vast, second only to the Pentagon. Mission focus was world-wide, with standby Army forces ready to respond to any emergency.

As usual, the first order of business was family housing. This time, it was not waiting for us. Fort McPherson had limited family housing and there was a waiting list. We lived in the installation guest house until we found a place to live. Fortunately, an old Army friend, Brig. Gen. Hank Hagwood, rescued us when he loaned us his second car. Also an Army aviator, Hagwood helped us become oriented to our new surroundings. Atlanta is a huge, sprawling, metro area. The car was a huge help while house hunting. We were able to rent a fairly new duplex condominium unit in the outlying city of Marietta, Ga., a 50-mile roundtrip commute in heavy traffic.

FORSCOM staff activity involved daily contact with the Pentagon. We had frequent coordination with 18th Airborne Corps Headquarters at Fort Bragg in Liberty, N.C. Rapid deployment forces there were always on standby, ready to respond to world-wide contingencies. The classified intelligence briefings I eventually received were a rude awakening to the grim realities of the dangerous world we live in. I often thought how the general public would have trouble sleeping if they were aware of what I learned.

Routine aviation staff section duties were fairly consistent and focused primarily on peacetime training and logistics. We conducted aviation resource management surveys, a fancy title more commonly known as an inspection. A full-time Army aviation experts team conducted the surveys at aviation units throughout the U.S. I was responsible for this program and usually joined the team for the results out-briefing to the local chain of command on the survey's last day. This program created a common focus on the key indicators of successful resource management throughout the aviation community.

My personal duties also included frequent briefings and presentations regarding Army aviation issues. My immediate boss was a one-star general who was the Assistant DCSOPS. A few months after my arrival, he was reassigned, and the Army did not assign a replacement. I now reported directly to two-star General Will Roosma. In August 1987, an important emergency arose. Iranian military forces attacked and interdicted flagged U.S. oil convoys exiting the Persian Gulf. The U.S. Navy provided protection. When the Iranians began to use small, missile equipped boats at night, armed U.S. helicopters were needed. Navy helicopter crews were not equipped and trained with night vision goggles, so the Army was called upon to assist.

FORSCOM immediately turned to the elite Task Force 160 Special Operations unit. Known as the Night Stalkers, this unit used the Vietnam-era OH-6 Little Bird helicopters, armed and fully night-vision capable. They are also extremely agile. Experienced and highly trained crews were ready to deploy worldwide on short notice. At the time, Task Force 160 was the only Army unit with these capabilities. With the operation code name Prime Chance, the Night Stalkers were promptly deployed. With no Army bases in the area, the Army used necessary unorthodox entry and operating means. Delivery aircraft secretly landed at night on a classified airport location where the OH-6 Little Birds were quickly unloaded and prepared for flight. Only two aircraft at a time could be inserted this way. They were flown to an oil barge platform which became their operation home.

Shortly after arrival, night surveillance began. It wasn't long before a missile boat was detected approaching a moving U.S. oil tanker. As this took place, we had a real-time communications link to the scene and monitored each piece of the action. When the target was confirmed, the Little Birds were cleared to attack. We sat nervously in the operations center in Atlanta and looked at each other while the first shots were fired. We wondered if this was the beginning of World War Three.

Sound familiar? This was nearly 40 years ago!

The target was destroyed, and things quieted down in a hurry. Task Force 160 remained on-station to continue their mission. We knew their ability to stay on location was limited. Already, the military was aware of the world-wide need for this capability. Normally, the Task Force 160 mission was short-term. FORSCOM was directed to develop a plan to replace Task Force 160 with helicopters that could remain indefinitely along with no mission interruption or degradation. This was a tall order. How could we replace the most highly skilled Army aviation unit quickly with something we didn't own?

I had to dig deep into my toolbox to find the answer.

Meanwhile, unexpectedly, I was informed by my boss that I would be moved into the vacant brigadier general position. I would assume direct supervision of all DCSOPS planning. This decision included assignment to general officer family housing at Fort McPherson where I would be immediately assessable to headquarters. I assumed this decision reflected the opinion among my superiors that I might be selected for general officer promotion when the board met in 1988. This raised new expectations which I had no time to dwell upon. I had my hands full with the tasks at hand.

The DCSOPS directorate included five separate divisions, each headed by a colonel. My former colleagues now reported to me. Fortunately, we shared a solid working relationship with each other and blended smoothly as a team. This was fortuitous because soon, another sensitive classified mission landed in our lap. We kept our fingers crossed for continuing success in the Persian Gulf when a new emergency erupted in Panama.

Self-imposed dictator Manuel Noriega had become a growing problem to the U.S. due to his drug trafficking involvement and other crimes including spy activity. The U.S. State Department determined it was time for a regime change and the removal of Noriega from Panama. FORSCOM was given primary responsibility to develop the operational plan to accomplish this. Fortunately, General Roosma had served in Panama. He had intimate knowledge of all the island and terrain. Astutely, he guided our thinking and analysis of how to best accomplish the mission.

Meanwhile, we still had the pressing need to do something about the Persian Gulf mission. I proposed the formation of a unit equipped with a new helicopter known as the OH-58D Kiowa. It was not as small and agile as the OH-6 but was equipped with the latest high technology operating systems. Its capabilities included carrying armaments such as missiles. These aircraft were just coming off the assembly line and their armament capabilities were being evaluated. I recommended the Army create a new unit at Fort

Bragg as part of the 18th Airborne Corps Aviation Brigade. It was commanded by my old friend and comrade Col. Ken Chien. After serving together in Vietnam and Hawaii, I knew Chien was the perfect leader to oversee this particularly challenging mission. The 18th Airborne Corps already had a Middle East orientation for standby missions with established lines of communication and logistics. All the pieces fit together.

After FORSCOM approved this plan, the next step was to brief the 18th Corps three-star commanding general, his deputy, and chief of staff. I went to Fort Bragg to do this and met alone with the three generals. When I completed my presentation, the commander general asked only one question. He wondered if we were actually serious about doing this. Yes, the concept sounded a little far out, but I assured him we were completely serious. Time was of the essence. After further discussion, they agreed their headquarters was best suited to manage this responsibility.

This was not a business-as-usual mission. There were major obstacles to overcome. It felt like mission impossible to form a new unit from scratch with a new aircraft and systems, train pilots and crews, plan and coordinate operational requirements all in a shorter than desired time frame.

My next step was to gain approval from the Army chief of staff to implement the plan. Guidance from the Department of the

Army was specific that our plan must ensure no degradation of support currently provided by Task Force 160. This was so easy for them to say. I was tasked to invent a unit to replace capabilities provided by the most highly skilled helicopter unit in the U.S. Army. And do it fast!

I was scheduled for a solo trip to the Pentagon where I would present a briefing. The plan was still held remarkably close, with only a few people aware of this huge project. I worked diligently on the briefing to ensure every concern was anticipated and addressed.

When I arrived at the Pentagon, I met with the Vice Chief of Staff of the Army General Art Brown. This is the same General Brown who had been my boss ten years earlier in Hawaii. His rise to four-star General rank had not diminished his calm, confident, and down-to-earth manner. As always, it was a pleasure talking with him. He wished me well with the briefing.

I entered the private conference room where Chief of Staff General Vuono and his immediate staff sat at a round table. Seated next to the chief, I used deskside flip charts for the briefing. I was the only non-general at the table. I was nervous yet fully confident with my presentation. Carefully, I went through each chart. There were no interruptions or questions. Occasionally, the chief responded with a nod of approval.

At the conclusion of the presentation, he looked across the table and directed a question to General Norman Schwarzkopf, Deputy Chief of Staff for Operations. He said, "What do you think, Norm?

General Schwarzkopf replied, "Looks good to me, Chief."

At that moment, my heart was thumping. My exhalation was audible. Everyone else nodded and that was it. The chief gave his okay and reminded me not to screw it up.

The details of what followed are many. Most importantly, it all worked and succeeded. The replacement helicopters were secretly air transported and delivered at night as the replaced aircraft were flown. To my knowledge, the Iranians did not detect our change of aircraft. Colonel Chien and his troops accomplished the impossible and there was no support degradation.

Throughout the planning and execution of Prime Chance and Just Cause, all communications were safe-guarded. It was during the early age of mobile phones. The ones we used were called brick phones because they were so large. They were capable of sending and receiving coded voice messages. I carried one with me 24-hours-a-day, 7 days-a-week.

The Just Cause full planning process was complicated. It took time to pull together all the pieces. The U.S. State Department felt a growing urgency to get Noriega removed. They put intense pressure on the Army to get it done. On two separate occasions, the Army requested to delay the mission because things weren't ready. General Colin Powell was President Reagan's National Security Advisor. Fortunately, he convinced the President to hold off. This was a joint air, ground, and sea operation with intricate timing elements for each participating element. I traveled to Central America to coordinate with the U.S. Army Southern Command. The trip's purpose was classified and planned accordingly.

My first stop was in Honduras where I visited a U.S. headquarters known as Joint Task Force Bravo. They oversaw long-term support for a major engineering project and humanitarian assistance. FORSCOM rotated active and reserve component units to this mission. They also had aviation support aircraft. My visit there was a cover for the rest of my mission.

In my next step, I wore civilian clothes and was taken by helicopter to El Salvador for a meeting to discuss the upcoming operation. This was tied to efforts inside Panama with the goal to gain local operation support. After we landed, I was met by a local person who said he would transport me to the meeting site.

He led me to a vehicle where the driver and an apparent security person waited. The security person had an Uzi machine gun on each arm. Surprised by this, I asked the guy in charge why the security person was armed. I was told not to worry. It was only a precaution. I asked, "What for?"

My handler explained that previously, the vehicle had been fired upon by Sandinista shooters, but they weren't expecting trouble that day. He assured me there was nothing to worry about. He added that the vehicle had armor protection. Unaware of this situation until I found myself in it, it was too late to turn back. Fortunately, we arrived safely without incident. I breathed a sigh of relief.

After the initial meeting, the officer I met with and I flew to Panama in an Army airplane where we continued coordination. Leaders inside the Panamanian Defense Force were anxious to see Noriega removed and were willing to help. There was no shortage of Panamanians anxious for Noriega to go. When I returned to Atlanta, I felt confident the plan was coming together.

Approximately two weeks later, we received word that the very same vehicle I rode in had been attacked and there were casualties. I breathed a prayer of thankfulness and again, realized that timing is everything.

Due to ongoing legal proceedings against Noriega, Just Cause went on hold. The invasion was delayed by a year. By then, President George H. W. Bush had succeeded President Reagan. He commenced the invasion on December 20, 1989. Considering the extreme difficulty of conducting a raid into a city environment to accomplish the mission, it was a considerable success. A total of 27,000 military personnel were involved including special forces, rangers, airborne infantry, ground infantry, and helicopters. Although casualties were held to a minimum, 27 U.S. military and three U.S. civilians were killed. Noriega initially eluded capture. He sought refuge but surrendered on January 3, 1990. He was transported to the U.S. where he was tried, convicted, and sentenced to 40 years in prison.

During my experience at FORSCOM, I gained a unique understanding of what goes into military contingency planning. As a result, I can fully appreciate what is likely going on behind the scenes. From my experience, there is a lot more to the story than what is evident in the mass media. Our military forces are always ready to respond with force whenever decisions are made to do so. I've provided a couple examples of continuous efforts in military operations centers that encapsulate planning and decision making from the Pentagon and with other headquarters throughout the U.S. military force. Many of these operations are 24-hours-a-day, 7 days-a-week.

While the world dangers kept us busy in 1988, another big issue was in play at FORSCOM. The commanding general led an effort to seriously address the defense within the continental U.S. He worked hard to convince the Washington D.C. establishment that we urgently needed to become initiative-taking about protecting our own borders. Unfortunately, his warnings were not taken seriously, and all attention remained directed at global scenarios. Diligently, he fought to transform FORSCOM into a joint headquarters that would undertake multi-service planning for the threats he foresaw. General Palastra was a man ahead of his time. The very case he tried to make 40 years ago is unfolding today. Despite the formation of the Department of Homeland Security following 9/11, we still have a serious border problem with no dedicated military role.

## A Moment of Disappointment

With a lull in action, in August 1988, we were able to get away for a week of vacation. We traveled to Wisconsin for a visit and rest. While in Wisconsin visiting my parents, I received a phone call from my secretary in Atlanta. She apologized for bothering me on vacation, but she had information I needed to hear. The brigadier general's selection list had just been released. My name was not on it.

Immediately, I felt a deep sense of disappointment. I regretted that I had assumed a false sense of hope that I would be chosen. After a few minutes of silent thought, I became honest with myself. This was not the end of the world. It was another one of those moments when I was reminded that certain things in life are meant to be. There are so many things we can't control.

I knew my official service record provided to the selection board was not as strong as others being considered. Two major facts stood out. I had never served at the Pentagon. At 50 years old, I was at the upper age limit of those selected. I believe the average age that year was 47. Should I have taken the Pentagon assignment I was offered after the War College? Maybe. But I still wouldn't have been any younger.

Not long after I returned to Atlanta, I received phone calls from mentors who expressed disappointment that I wasn't selected. They urged me to wait one more year. I was offered an important Pentagon job. I appreciated the confidence from these generals who knew me so well. I thanked them. Deep down, I had already accepted that my roller coaster ride had ended. I had no room for regret. Quickly, the Army assigned a brigadier general to the position I had temporarily filled.

I still had three years to serve before my mandatory 30-year retirement date. I decided that I would continue my service. I thought it might be best to seek an entirely new setting for my final assignment. I needed to begin the transition process for what I would do after I retired from the Army. I was young enough to do something else when I retired but had no idea what that would be.

Suddenly, another bugle sounded, just as it had so many times before.

An old aviator friend, Colonel Bill Sorensen, who served as commander of the Fort McCoy Garrison in Wisconsin, was retiring. I was struck by the idea that Fort McCoy would be a wonderful place to conclude my service; right where it began. This idea quickly took hold. I called Sorensen. He shared with me that his retirement would be in November 1988. He urged me to consider taking his place. Several things were underway at Fort McCoy of which I was not aware. Bill was a creative forward-thinker. He had developed the first-ever comprehensive plan for future modernization and growth of Fort McCoy's capabilities. It sounded exciting.

As I thought more about this opportunity, I became mindful of the many years we had been separated from our Wisconsin roots. My parents lived fifty miles from Fort McCoy. Jackie's brother and his wife were still in the area. The opportunity to spend quality time together was very appealing. The decision to go there began to feel like something we needed to do.

I spoke to my bosses and explained I would like to be reassigned to command the Fort McCoy Garrison. Initially, they questioned my motives and whether I was overreacting to being passed over for promotion. They urged me to wait another year. When I explained the details of my decision, they understood my reasoning and agreed to support me. The general in charge of personnel called Washington D.C. A day later, I was notified that reassignment orders were being published.

In 1988, most of the FORSCOM staff considered Fort McCoy a small, remotely located installation that was primarily a place for reserve component units to train in the summer. Budget wise, it was below the radar screen with exceptionally low priority. Nonetheless, it was still an installation that reported directly to FORSCOM. I began to look forward to the opportunity to get Fort McCoy on the radar screen.

When the word got out that I was leaving, each of the five colonels I supervised approached me privately to discuss my new assignment. They all asked what happened and seemed to think I must have really screwed up to be exiled to such an unimportant job. I tried to explain that I had requested the assignment. This was not a punishment. They were a great officer group that I thoroughly enjoyed working with. We shared a keen sense of teamwork and mutual respect. I was incredibly grateful for their dedicated efforts through each of the challenges we had faced. The concern they expressed for me was sincere and well intentioned. I would miss their friendship.

Jackie and I began to plan for one more Army move. We decided to vacate the family quarters so they would be available for the new General when he arrived. We stayed in temporary guest quarters until we left. After all the years of uncertainty and stress while waiting for what would happen next in the Army, we felt a sense of relief. This was over. It was a good feeling to know this was our last Army move. We had chosen it. We were happy.

# Chapter 18

# The Last Parade

I don't know of another soldier who had the opportunity to return to their first Army post and serve as its commander. Memories from my first night at the earlier Camp McCoy in July 1954 were still clear to me as I assumed the Commander's role in November 1988. The drive from Atlanta to Wisconsin was also nostalgic. We had made the journey from southern Alabama to Wisconsin many times in days gone by. This time, there was no denying that it was a sentimental journey.

As each mile passed during the long ride to Wisconsin, my thoughts wandered back over all our years as a military family. I wondered what our life would have been like had we never left Wisconsin.

What does this traveled road mean to the Boland family now? Occasionally, I've been asked, "What was it like, and how do you feel about it?" I have also been asked how being in the military affected Jackie and the boys.

My answers to these questions have surprised some who asked because overall, our feelings are positive. Despite the many hardships of constantly moving and starting over again, each of us agreed we wouldn't have traded those experiences for anything.

The boys felt that learning to adapt to new surroundings and people was extremely valuable. Jackie felt the many friends she made along the way are priceless. I agree. She amassed a collection of recipes that will be handed down for generations. She loved to cook and became an international gourmet chef. Everyone loved to be invited for dinner at Jackie's house.

We gained a special appreciation for what things in life are important and not to worry about things that aren't important. The boys gained a deep appreciation for equality among people. They grew up attending schools and having friends of diverse ethnicities and backgrounds. Military life builds a sense of common purpose among everyone. The idea that "we are all in this together" was very real and appreciated. The way we shared our unique experiences as a family forged an extraordinarily strong bond that will always be there.

Now, as we returned home, I realized we were blessed with lifetime experiences that most families will never know. Our special memories will always be with us.

Our family and old friends were happy to see us again. Soon it felt like we had never left. It was easy enough to talk about the past but more difficult to relate to the years in between. Military life is not very well understood by those who have not experienced it.

Similarly, we lacked appreciation and understanding of what it was like to never leave home. It took a while for us to fully reconnect.

When I arrived at the main gate entrance to Fort McCoy, it looked like I remembered it. However, I was surprised there were no military police guarding the entrance. It was wide open for anyone to enter. This seemed very strange compared to the 24-hour installation security I was used to everywhere else I had served. I expected Fort McCoy to be relatively quiet. This was too quiet.

Fort McCoy has a long history supporting U.S. Army training and mobilization. Founded in 1909 by Major General Robert B. McCoy, it is the only post named after its founder. Originally, McCoy acquired enough land to provide a training area for National Guard horse drawn artillery. The federal government acquired additional land and built a cantonment area large enough to support an entire combat division. During WW II, 35,000 soldiers were housed there. An excellent detailed history is found in a book authored by Linda Fournier, a former Fort McCoy employee.

Every building at Fort McCoy was old. The only modern building on the post was the Wisconsin State Patrol Academy facility where state and local law enforcement officers from across Wisconsin were trained. There was no family housing except for a handful of old buildings containing units considered sub-standard. Initially, we stayed in a transient guest unit housed in a manufactured home near the house where we would eventually live. We agreed to occupy a building once used as a cadre headquarters that was converted into a family home. After some renovations, we moved in. This became a wonderfully comfortable place to live and entertain guests.

Outgoing Commander Bill Sorensen provided me a full rundown of the Fort McCoy mission. I was extremely impressed with the planning he and his staff had completed for future improvements. Sorensen faced a major challenge in acquiring funding for each of these desperately needed projects. I inherited the same obstacle. Army models for infrastructure funding at its bases were determined primarily by the number of soldiers permanently assigned to the base. Fort McCoy had only 350 full-time soldiers. Most bases had thousands of permanent soldiers. At Fort McCoy, thousands visited for their brief training periods and were referred to as transient.

The issue became developing a rationale to justify facilities that reserve component soldiers needed and deserved. A modern vehicle washing facility was urgently needed. The plan included a state-of-the-art drive through washing center. Meanwhile, soldiers spent excessive amounts of precious little training time washing vehicles with garden hoses and scrub brushes before they could go home.

Sorensen successfully obtained funding for some projects, such as the washing facility. Other projects were still pending as they were stuck in the annual processes of construction reviews and funding decisions.

The many proposed projects included 80-units of family housing, a commissary (which the base lacked), a new post exchange, and a community club facility with a bowling center. Until these projects and numerous others became a reality, Fort McCoy would remain locked in the 1942 time period. The master plan provided a blueprint and path for a significant step forward. While the plan was ambitious, the improvements were desperately needed.

Fortunately, the old wooden buildings were in surprisingly good condition and still useable for temporary housing. Minor construction renovation projects improved the serviceability of some barracks, especially for cold weather occupation. Conversion of the barrack's heating systems to propane furnaces was an example. Not long after my arrival, winter snow blanketed the base, and I observed signs of installation security lapses. Approximately half the barracks buildings still used the original coal furnaces with a concrete coal bin adjacent to the barracks. On several occasions, I observed vehicle tracks leading to coal bins where it was evident coal had been removed. I wondered what else might be going on that shouldn't be.

My early impressions confirmed the image Fort McCoy had at the U.S. Forces Command Headquarters (FORSCOM): antiquated facilities that were "good enough for reservists." No one believed there was a particular need to make improvements to the base. No Army decision makers cared about the Fort McCoy situation. While Fort McCoy had been designated a fort in 1973, the facilities had not been modernized.

From a combat training standpoint, Fort McCoy had much to offer. But without a sizable number of full-time soldiers, it was doomed to remain a low priority for modernization. I could have adopted this same attitude. This was a perfect place for me to relax, wind-down from the fast pace I had lived for much of my military career and prepare for life after the Army. This seemed to be an unlikely place where any unusual or classified missions might take place. But I wasn't at all inclined to assume a retired on-active-duty attitude. I still had energy, enthusiasm, valuable experience, and the desire to serve our country until the last day of my career. I couldn't see a clear path for action and wondered how I would proceed. Shortly, this drastically changed. The bugle sounded again, just as it always had.

## Job One – Keep Fort McCoy Open

The first major departure from ordinary procedure was base realignment and closure (BRAC). With growing pressure to reduce base operating costs, Congress directed the creation of an independent commission to study all military bases and make recommendations for closures and consolidation; a realignment of missions. Anticipated huge cost savings were expected by closing installations that contributed little to national defense requirements and little perceived value. When this study was announced, it didn't take long for a sense of panic to set in among those who cared about Fort McCoy, especially the local communities. Fort McCoy is situated halfway between the cities of Sparta and Tomah, Wis. The base had approximately 1,600 civilian employees that reside in these communities. Given the obvious low priority the Army had shown for Fort McCoy with its limited year round use, it was logical to assume the base was a prime candidate for closure. BRAC became headline news in Western Wisconsin. My phone rang off the hook with calls from concerned citizens, elected officials, and local media. My short honeymoon was over.

My first step was to get my arms around what the study entailed. I needed to become familiar with the criteria used to evaluate each installation. After I realized the evaluation's entire scope and details, I was not only surprised, but overjoyed.

The plan included a very comprehensive and detailed scoring system designed to determine military value. Facilities that were capable of supporting combat training and firing of all Army weapons were most heavily weighted. My immediate reaction was, "Welcome to Fort McCoy: we can do it all!" I had served or visited most U.S. Army installations and was familiar with each installation's capabilities. The BRAC plan indicated that installations located in urban areas with limited training acreage were least valuable. I immediately thought of nearby Fort Sheridan in Illinois. With a longtime country club setting, it is nestled on Lake Michigan's shore. Traditionally, it was home to administrative missions. The base included many historic buildings and a golf course but no training areas. Soldiers used a brand-new commissary and post office. The base was commanded by a three-star general. Several friends assigned there insisted there was no way Fort Sheridan would be closed. They indicated that Fort McCoy was toast.

My only concern was whether the commission would strictly adhere to the designated scoring system or if there might be any hidden tricks that would skew the results. There was intense political pressure among members of Congress who fought fiercely to prevent an installation in their district from being closed. In light of this, the study's lawful requirement stated that it had to be accepted by Congress in its entirety or not at all. If the commission stuck with the script, I was confident Fort McCoy would survive and rank high on the retention list.

Fort McCoy also had a standby mission to support a national mobilization of reserve component troops. Most of its capacity was unused. When I was called up in 1961, Fort McCoy was not utilized. While there was nothing on the horizon that suggested a pending mobilization, training exercises were occasionally conducted to practice for this possibility.

## Early Fort McCoy Days

Despite the upswing in attention after the BRAC evaluation, I still had ample time for family activities. Everyone enjoyed visiting the old house we lived in. It became a tradition for my father to ceremoniously light a fire in the huge fireplace. We especially enjoyed the popping logs and how a fire warmed the house and our souls during the holiday season.

My father had a life-long desire to visit Ireland. He had no connection to anyone there but simply wanted to see where his ancestors came from. In November 1989, his dream come true. Diagnosed with esophageal cancer in 1985, my father underwent a laryngectomy. Afterwards, he diligently learned to speak audibly. Now 74 and in overall good health, our wives gave us clearance to make the trip without them. Away we went – just the two of us.

We spent an amazing week together. Quickly, we found out

driving a rental car in Ireland was more stressful than we anticipated because everything is backwards from driving in the U.S. Navigating a steering wheel on the right and driving on the left side of the narrow roads became a harrowing experience. Operating the manual shift with the left hand was an additional wrinkle and added to the fun.

Before we arrived at our first intended destination in Galway, Dad announced he could already see the countryside was very green. He recommended we cut back on our planned sightseeing and spend more time experiencing the local culture by visiting pubs. Did I mention how much my dad enjoyed beer? With our revised plan, we consumed plenty of culture. We made sure to kiss the Blarney Stone for good luck and especially enjoyed our time in Killarney. Memories from our special week together still warm my heart. I can see the happy look on his face singing *When Irish Eyes Are Smiling*. Until the day he died in 1995, Dad claimed this trip was his fondest memory.

## Making the Case for Fort McCoy

After I left FORSCOM, General Colin Powell became the commander and was my new boss. During the 1989 summer training period, he made a full day visit to Fort McCoy. It was such an honorable and memorable experience to meet and get to know the most respected officer, statesperson, and individual in the Army during my era. Powell was very pleasant and easy to talk with. He left an extremely strong lasting impression. I provided him with an overview briefing about Fort McCoy, which, I must admit, included a hard sell on Fort McCoy's potential. I thought I was quite clever as I used the BRAC criteria to justify my claims. I pointed out that Fort McCoy would be an excellent place to permanently station active Army combat troops. I advocated that the Pentagon make the investments to transform Fort McCoy from a WW II camp into a modern Army fort.

When I finished my presentation, Powell looked me in the eye. His face widened with a broad grin which caused me to think I had nailed the briefing. Then General Powell said, "Ray, it's interesting how every installation I visit, I hear how it's the center of the universe."

I gulped and said, "Sir, at Fort McCoy, that may actually be true!"

He smiled again and said the Army would continue to support the currently approved projects, but the rest would require more study. I offered to provide his staff a more detailed proposal. I was satisfied that we may have opened the door to a brighter future for Fort McCoy. Three months later, I was disappointed, but not surprised, when Powell was chosen as Chairperson of the Joint Chiefs of Staff. General Edwin Burba was named the new

FORSCOM commander. One year later, in August 1990, General Burba visited Fort McCoy.

Armed with my experience from briefing Powell, I refined the script for a similar briefing with Burba. This time, I included more specifics on how to grow Fort McCoy's mission and modernize its supporting infrastructure. By now, I was no longer reluctant to express my candid opinion to Army leadership, regardless of any political repercussions. It was clear that my best final contribution to the U.S. Army would be to advocate for Fort McCoy's ability to play a more defined key role in national defense preparedness. I described Fort McCoy as the Army's best kept secret.

Burba showed genuine interest in Fort McCoy's potential. He asked that I follow-up with a written proposal. Several large Army posts were so crowded that units had difficulty scheduling training areas and firing ranges. At Fort McCoy, these same facilities were unused most of the time. Just when this discussion with General Burba became interesting, it was quickly interrupted. General Burba's travel team informed him that he had an emergency call on their mobile tactical satellite phone. Iraq had just invaded Kuwait. Burba had to return to his headquarters in Atlanta immediately.

I wondered why Iraq would invade Kuwait. This was not a scenario that we had studied at FORSCOM. I wasn't sure what to make of it.

## Operation Desert Shield

Two weeks later, a scheduled mobilization training exercise took place at Fort McCoy. During this training, we received simulated mobilization messages from FORSCOM in a secure, controlled access operations center adjacent to my headquarters. Suddenly, one of the officers working the operation came to my office. We needed to speak immediately. The operations center had received messages from FORSCOM that were "real," which means President George H.W. Bush had ordered an emergency call-up of reserve component soldiers. Fort McCoy should be prepared for units to arrive soon and prepare them for overseas deployment.

This was the beginning of Operation Desert Shield. Quickly, things become more exciting at Fort McCoy than they had been in a long time. A prelude for the U.S.'s war time involvement in the Middle East, this development impacted Fort McCoy in a way I never expected. I never envisioned becoming the commander of the busiest mobilization site in the U.S. I no longer had time to think about the future. My hands and mind were filled beyond capacity with the present. Now, I lived a daily situation that resembled the good old days at FORSCOM with 24-hour-day, 7-days-a-week real world implications.

Many unforeseen challenges arose that were never uncovered and addressed during the pretend training exercises and plans. Units arrived at Fort McCoy with all their vehicles and equipment. The requirement to repaint every vehicle with the desert camouflage pattern was not in the plan. This must be accomplished before moving equipment by rail to the east coast for sea shipment to the Middle East. Fort McCoy had an extremely limited paint shop capability. Yet, we painted over 3,000 vehicles. What a big challenge! The required special paint was in short supply. Every mobilization site needed it. Only because of the amazing resourcefulness of our superb Director of Logistics Bill Klein was this job accomplished. To this day I don't know how he did it. Maybe it's best I don't.

Meeting the compressed schedules for each individual unit was the biggest challenge. The first soldiers arrived the end of August. Many more came in September. All deployments were scheduled for completion by January 1, 1991. We received detailed movement deadlines that could not be altered. Each rail movement had to arrive on time for sea shipments planned to arrive in the theater when the soldiers arrived by air. It was mind boggling. Some rail loading required all-night efforts. I recall some trains rolling out of Fort McCoy with wet paint running down the sides of equipment.

The Army has two tiers of planned mobilization staffing, known as partial and full mobilization staffing. Desert Shield was a partial mobilization which meant I was the senior commander responsible for the entire Fort McCoy mission. Under full mobilization, a two-star general takes command. As planned, Fort McCoy received additional staffing resources to augment our permanent staff. The primary element was the 5064th U.S. Army Garrison Headquarters, which was an Army Reserve unit from Detroit, Mich. Although they were structured to oversee the entire processing of the mobilized units, I completely merged their staff with mine to create a fully coordinated team. I appointed their commander, Colonel Al Gentry, to serve as my deputy. In each staff section, the senior ranking officer was placed in charge. Gentry and his people worked ridiculously hard with my personnel. Every bit of energy our combined staff had was necessary to keep up with everything.

My most valuable staff member was Fort McCoy's Command Sergeant Major Larry Martin. He was an amazing non-commissioned officer with an extraordinary background. After serving in the special forces, he was recruited to work for the CIA in Laos during the Vietnam War. He spent six years in Laos and was one of the last Americans to leave in 1975. Together, we mentored each unit commander and senior non-commissioned officer to help them confront the responsibilities they faced and gain confidence to succeed. We encouraged them to address things they thought were important within their units. These were often things reports and data don't necessarily show. These personal level interactions were terribly effective.

Another key player was the civilian Chief of Staff Al Fornier. While Gentry, Martin, and I were buried in daily mobilization activities, someone needed to oversee the normal Fort McCoy mission activity. Much of this responsibility fell to Fornier. He was more familiar with the nuts and bolts of every installation function than anyone. His ability to work effectively across the staff functions and labor relations was essential to the civilian workforce's excellent performance. His many years of experience in his duties made him a true expert. I can't imagine how we would have kept everything going smoothly without him. He is an unsung hero in the long-term Fort McCoy legacy.

As each individual unit arrived and was processed, there were many cases of extreme hardship. Often, this reminded me of my own mobilization in 1961. The most common hardship was the sudden disruption of employment and reduction in household income. People from all walks of life were called to serve. There were few authorized deferment reasons. Doctors, lawyers, teachers, and first responders were suddenly uprooted from their families and communities. Another important family concern was health care. Some employers continued to provide existing coverage, but many didn't. We had to furnish active-duty identification cards to ensure military health care coverage. Fort McCoy had no full-time medical staffing. We received assistance with this from other active Army installations. Fort Leonard Wood staff came to Wisconsin and helped with medical and dental exams, which were required before soldiers left the U.S.

The list of issues was lengthy and exhausting. I've provided just a few examples. I was also blindsided by the unforeseen media avalanche that demanded information. It had been nearly 30 years since something like this happened. The fear of the unknown was very real. When I was called up in 1961, it became readily apparent that we weren't being deployed overseas. This call-up was completely different. There was no doubt that soldiers were heading overseas.

Desert Storm caused a high state of anxiety nationwide. In 1990, the whole purpose for overseas deployment was to face the forces led by a dictator described by some as the most dangerous man on earth. Saddam Hussein was known to have used chemical weapons. This alone was frightening. It became public that most units would deploy to Saudi Arabia. Once there, it was anyone's guess what would happen. This was a new experience for most Americans.

In more recent times, mobilization of reserves has become a normal event. Many soldiers are deployed more than once. I never dreamed the day would come when the weekend warriors of my earlier days in the military would become full-time partners with active-duty Army soldiers in meeting world-wide contingencies. Today, reserve units serve on active-duty at many overseas locations. We rarely hear of these situations. This is a significant and dramatic shift from how these units were previously managed.

Many mobilized units came from small cities and towns where their departure was felt throughout the community. Fort McCoy received units from the Dakotas to Ohio and eventually totaled 10,000 soldiers. It wasn't long before many family members began to arrive, along with various media outlets. At one point, five satellite television vans were parked near my headquarters, providing live feeds to their stations. None of this was foreseen in pre-mobilization planning. Quickly, we had to improvise. Their interests couldn't be ignored. During my communications studies, I learned that the more information I provided on a timely basis to the public, the better reception and coverage we would receive.

We converted an old service club recreation facility into a visitor center which restricted visitors from roaming around the post and talking with soldiers without our knowledge. We provided daily updates on soldier activity and anticipated deployments. We established a scheduled news conference three-days-a-week at 10 a.m. and urged every outlet to attend. Typically, 10 to 15 representatives from print and electronic outlets attended.

I personally conducted these updates and provided as much information as possible within security clearance. FORSCOM public affairs guidance was overly restrictive and unrealistic. I made up my own rules, knowing full-well what information was actually classified. My open approach established credibility and trust with everyone. They understood I was giving them my best shot. The question-and-answer sessions included much interest in the capabilities of the U.S. and Iraqi military. Fortunately, I had full knowledge of all the various possible weapons from my days in Germany. I was able to state confidently that the Iraq Army was no match for ours. My assessment proved accurate when we easily overpowered the overrated Iraqi forces.

The most requested information involved the days and times soldiers would leave. Most were flown out of the nearby Volk Field Air National Guard Base. I announced air departures within 72 hours of departure. Many family members went to Volk Field to say their final goodbyes. Thousands of donated cookies and cups of coffee were available. I visited with many family members. One requirement included that an Army band play at each departure. I became acquainted with many members of the 32nd Wisconsin Army National Guard Band. They were wonderful and showed up no matter the time of day or night they were needed. On one occasion, about 3,000 people were on hand for a departing unit.

In mid-October, about 2,000 soldiers were nearing their deployment date. Bob Hope was scheduled to perform at the La Crosse, Wis., Civic Center on October 16. Famous for his many years of entertaining troops, we wondered if there was any way he could perform for our soon-to-deploy soldiers. We went to work to make something happen. After coordinating with the La Crosse venue, we found out his schedule was very tight. There was no time to do anything other than his scheduled performance. Bob Hope wanted to include the soldiers in his performance. He offered to host the soldiers at the La Crosse performance at no cost. All we had to do was get them to the venue. We organized a convoy of 40 buses to transport the soldiers to the show. When the buses arrived at the arena, the soldiers formed up. Each unit marched into the arena carrying their unit guidon flag. The crowd cheered as they entered. The soldiers filled the arena's upper-level seating area. Effectively, this turned into a Bob Hope soldier show. Chief of Staff Fornier coordinated all the details. The rest of us just showed up.

Part of the plan called for me to meet with Bob prior to the show in his dressing room. As I sat alone with Bob, we reminisced about years gone by. I experienced his shows on two separate occasions in Vietnam and brought along a book from the 101st Division that included photos from his visit in 1970. He was happy to sign the book for me. As we talked, a warmup act performed.

While Ann Gillian sang *The Wind Beneath My Wings*, we heard the growing enthusiasm among the soldiers in appreciation for this special send off.

As Bob entered the arena and approached the stage positioned in the round, his famous orchestra played his theme song. As I went to my seat, the soldiers cheered for Bob. I noticed their voices HAD become ever more increasingly loud. Command Sergeant Major Martin, a stickler for attention to detail, assured me all the pre-function checks were completed before the event. However, one small detail was overlooked.

During their entire time at Fort McCoy, the soldiers were in lock-down. They were restricted to their barracks areas for completion of all required training and pre-deployment preparations. There was a no alcohol policy during this period. The arena had numerous concessions stands near the seating areas that served the largest paper cups of beer I had ever seen. By the time Bob Hope began his show, the troops were definitely warmed up. Bob appeared to appreciate the soldier's response and he became more enthusiastic as well. He was 87 years old and his performance was remarkable. He sang, danced, told many jokes and seemed to grow younger minute-by-minute. He never missed a beat. When Bob announced he would be going to Saudi Arabia for a Christmas show, he told our soldiers he looked forward to seeing them again.

As planned, I joined Bob on stage at the show's conclusion. I thanked him for a great send off for our soldiers and presented him with a desert camouflage Army jacket and matching hat. His name was sewn onto the jacket and included the words, *Fort McCoy*. Immediately, Bob wanted to wear it, so I helped him into it. Then, he proudly strutted about the stage.

I was holding a wireless microphone and gave in to a last second impulse. I said, "Bob, the soldiers asked me to tell you to be sure to wear this uniform at all times when you visit them, because they don't want you to get caught between a rock and a hard place." The crowd roared and Bob flashed a broad grin. The photo taken at this exact moment is my all-time favorite item of memorabilia. As the band broke into his famous theme song, *Thanks for the Memory*, he walked off stage. The audience continued to applaud until he was out of sight. As I exited the stage, I had goose bumps. What an amazing evening it had been.

But it wasn't quite over. Suddenly, the leader of Bob's staff approached me. He blurted out, "What's wrong with you!"

"Why?" I replied.

"NOBODY cracks a joke on Bob," he cried to me. "You don't do that. HE tells the jokes!"

I was riding a high of euphoria from the whole experience and was in no mood for this guy's condemnation. I looked at him and said, "Sorry, the devil made me do it."

I thought I had the last word, but this story is still not over. Two weeks later, I was in a meeting with senior staff when my secretary interrupted. She had a phone call I needed to take right away. Immediately, I wondered if it was the FORSCOM commanding general and asked, "Who is it?"

My secretary said Bob Hope was on the line. Smirking, my colleagues thought this was a joke. Before I could say anything else, she emphasized that it really was Bob. He wanted to talk to me.

I picked up the phone. Sure enough, it was him. "Hey Colonel, how are you?" he said.

"I'm great, Bob," I replied. "We are still talking about your show."

Bob shared that he really enjoyed the time with us, and he wanted to ask me a favor. I said, "Sure, what is it?"

He recalled that I cracked a joke at his expense but couldn't remember what it was. He asked if I would tell it again because he wanted to use it! (He was famous for using borrowed jokes.) I told it again and he thanked me. Then, I asked him for a favor. Would he sign a copy of the picture with the two of us? He instructed me to mail it to him and he would. We ended the conversation by wishing each other well.

When Bob visited the troops in Saudi Arabia, national television production teams recorded the performances that were also aired in the U.S. In every show, he wore the Fort McCoy jacket. During the show, he also told "our" joke. To my knowledge, these were his last Christmas soldier shows. His La Crosse show was his last performance in the U.S. Soon after these shows, at his wife's urging, Bob finally retired. He lived until he was 100. My October 1990 night with Bob Hope seems like a storybook ending. Maybe it is. Bob performed his first USO show entertaining troops in 1941. We both completed our service 50 years later. For me, it seems like the night in LaCosse was meant to be just like so many other parts of my story.

Desert Storm didn't last long. As anticipated, the U.S. led military effort easily defeated the Iraqi forces and sent them back to Bagdad. As spring of 1991 arrived, many units returned to Fort McCoy. We held a final ceremony for each unit before they went home and thanked them for their service. The post-mobilization phase continued for some time afterwards. The return and repair of equipment was called Desert Fix.

### Will Fort McCoy Stay Open?

As my July 1 retirement date approached rapidly, I realized there hadn't been time to prepare for life after the Army. Right up until my last week of service, I was still struggling to keep new construction projects moving forward under the cloud of pending BRAC announcements. Eighty family housing units were scheduled to be built in Tomah under a new build-to-lease program managed by a developer. The funding source had cold feet despite my assurances that Fort McCoy was not on the chopping block. During one of my last conversations with the developer, I was told, “Easy for you to say, but it’s OUR money.”

As we awaited the BRAC results, Wisconsin elected officials attempted to weigh in with a strategy aimed entirely at keeping Fort McCoy open. Maintaining status quo was the goal. I was in no position to enter the political arena and point out the opportunity to promote mission growth. In the end, Fort McCoy remained open. It ranked 23rd out of the 90 evaluated installations. Fort McCoy was in the top one-third along with the largest and most populated Army installations in the U.S. Fort Sheridan was closed. As a result of this evaluation, Fort McCoy was assigned responsibility to manage all installation equipment and furnishings disposition and disposal.

The official announcement that Fort McCoy would remain open happened just a few days before I retired. While this was good news, I felt time had run out for me when there was much more I wanted to accomplish. This was not a new feeling but one I had each time I moved on to the next Army assignment. This time, it was different. Now, it was time to move out.

The change of command ceremony took place on July 1, 1991. The presiding officer was Lt. Gen. James Hall. He was the Headquarters 4$^{th}$ U.S. Army Commanding Officer at Fort Sheridan. My immediate boss, he was a pleasure to work for during Desert Shield. Hall was an outstanding commander who placed his full trust and confidence in me and let me do my job. He was supportive when I needed help. I knew that I could always count on him. He gave us a wonderful and warm send-off during the ceremony. He presented Jackie an Army Certificate of Appreciation for her many years of service. I received a third award of the Legion of Merit.

The new commander was Colonel Bill Stanley. During the change of command ceremony, Stanley received the Fort McCoy flag. As the previous commander, it was my job to stand down. My service was officially over.

As is customary, a welcome reception was planned for Stanley with all the attending guests invited to the Officers' Club. After some handshakes and a parting television interview, only Jackie and I were left standing on the parade field. I still had a tear in my eye from when the last parade of soldiers passed. In the moment, I had felt an enormous sense of pride for the privilege of serving our great country for 30 years. It was an honor that I could never duplicate. I was mesmerized with these thoughts when I heard Jackie say, "Honey, it's time to go. It's over."

Just like that, my story ended.

## Discovering More Than a Story

I think it would be difficult for anyone to walk away from 50 years of amazing experiences and not wonder, "What did it all mean?" It is important for anyone who ponders this question to reach inward and discover what was most important in his or her life.

So, what did I learn? Were my military experiences important and valuable? Have I made a difference? Often, I have pondered these things. Here are my conclusions.

The early years of my life were mostly about growing up and trying to figure out who I was. This period included the very difficult process of becoming honest with myself. I had to find my true self instead of trying to be something I wasn't. This was a huge step. Until I confronted the reality of making peace with who I was, I could not begin to set meaningful goals.

Setting goals was another important step. I recall performing many interviews with soldiers who were considering re-enlistment before they were discharged. Often when I asked about their future plans, the response was, "Get a job." Some answered in such a way that demonstrated they had specific goals. In the early days, my personal goal was simply to support my family. It took time before I figured out that I wanted to serve my country as a soldier.

Once I had established this general goal, I discovered the meaning of the cliché, "The devil is in the details." A long winding road of twists and turns, obstacles, setbacks, and disappointments often questioned my ability to succeed. In those difficult and sometimes disappointing moments, I *had* to believe in myself and trust my instincts. This important life lesson is not always easy to embrace but is imperative.

This is a universal reality. Regardless of whatever someone tries to accomplish or become, similar challenges will be present. It seems to me, the pathway to achieving any goal is like climbing a ladder. The only way to get to the next rung is to reach for it. Yet, reaching for the next rung is not a requirement. It is a choice.

Many people are very happy exactly where they are in life. They have no desire to climb higher. If this is you, please celebrate this. Yet, making choices is a never-ending challenge. With choices come mistakes. Often, I had to remind myself that today was a new day. It helps when we learn from our mistakes and try not repeat them.

My most important life achievement has been to find happiness. Success and happiness go hand-in-hand and are always in the eyes of the beholder. Individually, we must define what happiness represents for us. Measurements are not necessarily material. There are people who are wealthy yet not very happy.

Others, with modest means, are very happy. The ladder of life should lead us to what makes us happy. I encourage you to choose what makes you happy, rather than what society, culture or someone else feels is best for you.

At various intersections of my life, I faced multiple questions: Who am I? Where am I going? How do I get there? What will make me happy when I arrive? I chose to face these questions and, fortunately, found the answers. Faith, family, and friends showed me the way. Hard work and perseverance helped me cross the finish line. This formula can work for anyone, regardless of their goal. While goals are what we want them to be, the following advice is worth noting:

> *"The greatest danger for most of us is not that our aim is too high and we miss it, but that it is too low and we reach it."*
>
> - *Machiavelli*

## The Faint, Distant Bugle

After the change of command at Fort McCoy ended, Jackie and I walked silently to the parking area. I paused when I heard the faint call of a bugle in the distance. For now, I walked off the field knowing so many bugle calls had been summoned and completed in these years of my life. Bugle calls that took me to geographical places I never anticipated visiting. Experiences that stretched me beyond my imagination. Opportunities that taught me so much about myself. Faith that I could accomplish more than I ever imagined. Gratitude my family willingly accepted the sometimes difficult calls and lived them out with me, often from afar. The feelings. The emotions. I could barely comprehend it all.

Maybe the bugle call would sound once again.

That's another story for a different day.

# Epilogue

Thank you for reading my story. Hopefully, these words have touched you in some small way and you connected to some part of my journey.

As I worked through what to leave in and what to let fall to the editing floor, I began to see three distinct phases of my life. The earlier chapters recall my growing up years. As I recalled this phase of my life, I realized how immature I was in my early years. While there are certain things a person discovers purely with experience, it may have taken me a bit longer to discover how much I needed to grow up. I am thankful for the various mentors who didn't give up on me when I made mistakes. They shaped me and molded me so that when I reached the next season of my life, I was prepared.

In the second phase of my life, I discovered that I really could become good at something. I became a solid soldier and contributed to the U.S. military. With proper training and another round of helpful guides and coaches, I discovered that I fit into the business of the U.S. Army and larger military picture. Finally, I hit a stride with my life. Because there were things with which I was actually gifted, I made a difference.

My next life responsibility challenged me to use the gifts and talents at which I was good. Honestly, no one was more amazed by this revelation than me. With maturity and experience now safely part of my toolbox, I discovered the capacity to accomplish some really good things: for myself, the benefit of others and the U.S. Army.

I would be remiss if I did not include the most important ingredient that overarches my entire life. As an 86-year-old person glancing into the rearview mirror of my life, I have the definitive and strong belief that God has a plan for each of us. The only explanation I have for why so many certain things happened in my life can only be attributed to God's divine intervention and design. I have felt God's hand and Spirit guide me thousands of times, often on a daily basis. Countless times, I found myself physically, emotionally and mentally drained and depleted. In these moments, I relied upon the spiritual being I call God to take the wheel of my life and keep me on the path.

There are many, many things that we cannot control in this life. Yet, we have a say as well. God's plan is not a blueprint with every detail already accounted for. We get to partner with God and work out the details together. You and I have an individual responsibility to find the destinies that await us. Part of our life mission is to consult and look for inspiration from God as we make big and small decisions.

Seeking and implementing this plan will not simply be dumped into our laps. This journey is not intended to be easy. We will be bombarded with constant choices and decisions. Sometimes, we will make a more correct decision. Other times, we may find ourselves in a position where we wished we could have a do-over. Mistakes are part of the path. When we are knocked down with disappointment, only we can decide to lean into our faith, get up, dust ourselves off and try again. When tragedy strikes, and believe me, it will, we must be the ones to dig deep and find the faith to carry on.

Has my will been tested beyond my expectations? You bet! While I may not have fully realized it in the moment, a strong sense of faith is what enabled me to keep moving forward. Thankfully, a seed of faith had grown roots into my heart and soul in my early years. Even when I felt engulfed in a series of dark days or weeks or months, this tiny seed of faith gave me hope that the darkness would one day become a bit less daunting.

During the early years of my life, there were times when I had no idea how to take the next step. I had nothing but faith to sustain my will. I reassured myself that somehow, some way, if I kept trying, I would reach my next goal. Amazingly, it happened!

In the mid-years of my life, a series of unexpected obstacles and diversions seemed insurmountable to achieve success as an officer in the U.S. Army. I kept plodding. Sometimes it felt like I was going forward. Other times, I wondered if I spent more time in reverse. Eventually, things worked out. The pieces of the puzzle came together. As I felt more in-step with God's guidance, I knew that my gifts and talents were divinely ordained and would be used for God's kingdom and the betterment of humanity.

When I discovered how to use what I learned, I accomplished things that previously seemed impossible. I was blown away with the results! Finally, I discovered within me the will and faith to succeed. This success was not only for me.

Were the bugle calls always distinctive and clear? No. Did I question myself and my decisions along the way? Countless times. Would I have preferred more clarity at specific times? Absolutely. Did some of the bugle calls leave me in the wake of disappointment and disbelief? No doubt.

Yet, this is the story *of* life. More appropriately, the story of *my* life. As I have written these words and reflected upon my life journey, I am completely amazed how often God helped steer the direction of my life. Yes, I made the choices, but they were never completely in a vacuum.

Too often, people think if God is a part of his or her life, then things will be easy. God will cover us with a shroud of protection and keep us from evil and bad things. This is *not* how faith works.

Faith is what we depend upon to help us slug through the awful and difficult times in our lives. It is the safety net that catches us when we want to give up. My will was tested every step of the way. Yet, I embrace God's love in knowing that anything really important in life may not be easy. The big things are worth the effort.

Today, I am completely at peace with where God's plan took me. These days, I find comfort in knowing that God is always with me. I have no reason to fear evil or anything else because the One who has always been my co-pilot continues to fly with me today.

# Acknowledgments

My Mother for giving me a strong sense of determination.

My Father for giving me a strong sense of fairness and compassion.

Master Sergeant James Mancini for teaching me the meaning of leadership.

Ida Connery, my high school English teacher, who challenged me to write.

Captain Ray Bartkowiak who encouraged me to enter officer candidate school.

Lieutenant Colonel Frank Henry, my strongest leadership role model.

Major General James Smith who gave me the opportunity that changed everything.

Toros, B Battery, 4/77 ARA

Reliables. 18th Aviation Company

General Arthur Brown for mentoring me on senior officer responsibilities.

My son David for helping me in many ways.

Acknowledgements

Dianne Vielhuber for her patience during the book editing.

My wife, Donna, for her unwavering support and encouragement to finish the book.

# Acronyms

| **Acronym** | **Meaning** |
|---|---|
| 1 LT | First Lieutenant |
| 2 LT | Second Lieutenant |
| AC | Active Component |
| ADAO | Assistant Division Aviation Officer |
| ADC | Assistant Division Commander |
| ADF | Automatic Direction Finder |
| ADIZ | Air Defense Identification Zone |
| AH-1G | Attack Helicopter – Cobra |
| AO | Area of Operations |
| AOC | Advanced Officers Course |
| ARA | Aerial Rocket Artillery |
| ARVN | Army of the Republic of Vietnam |
| ATT | Army Training Test |
| AWC | Army War College |
| BC | Battery Commander |
| BG | Brigadier General (One-Star) |

## Acronyms

| | |
|---|---|
| BRAC | Base Realignment and Closure |
| CCN | Command and Control North |
| CG | Commanding General |
| CGSC | Command and General Staff College |
| CH-47 | Chinook Army Helicopter |
| CO | Commanding Officer |
| COL | Colonel |
| CP | Command Post |
| CPT | Captain |
| CSM | Command Sergeant Major |
| DCSOPS | Deputy Chief of Staff for Operations |
| DIV ARTY | Division Artillery |
| DMZ | Demilitarized Zone |
| E-4 | Specialist/Corporal |
| E-5 | Sergeant |
| E-6 | Staff Sergeant |
| E-7 | Sergeant First Class |
| E-8 | Master Sergeant/First Sergeant |
| E-9 | Sergeant Major |

| | |
|---|---|
| EXO | Executive Officer |
| FARRP | Forward Area Refuel & Rearm Point |
| FDC | Fire Direction Center |
| FEB | Flight Evaluation Board |
| FO | Forward Observer |
| FORSCOM | U.S. Forces Command Headquarters |
| GEN | General |
| Grid Coordinates | Tactical Map Location |
| HE | High Explosive |
| HQDA | Headquarters Department of the Army |
| ID | Infantry Division |
| IFR | Instrument Flight Rules |
| IG | Inspector General |
| IP | Instructor Pilot |
| JAG | Judge Advocate General |
| KP | Kitchen Police/Patrol |
| LTC | Lieutenant Colonel |
| LTG | Lieutenant General (Three-Star) |
| LZ | Landing Zone |

Acronyms

| | |
|---|---|
| MACV | Military Assistance Command, Vietnam |
| MAJ | Major |
| MARS | Military Auxiliary Radio System |
| Med Evac | Helicopter Medical Evacuation |
| Mess Hall | Dining Facility |
| MG | Master General (Two-Star) |
| MIA | Missing in Action |
| Mini-Gun | Multi-Barrell Electronic Machine Gun |
| MP | Military Police |
| MSGT | Master Sergeant |
| NATO | North Atlantic Treaty Organization |
| NCO | Non-Commissioned Officer |
| NG | National Guard |
| NOE | Nap of the Earth |
| NVA | North Vietnam Army |
| O-1-A (L-19) | Bird Dog (Airplane) |
| OBC | Officer Basic Course |
| OCLL | Office of Congressional Legislative Liaison |
| OCS | Officer Candidate School |

| | |
|---|---|
| OER | Officer Efficiency Report |
| OH-13 | Observation Helicopter – Sioux |
| OH-58A | Observation Helicopter – Kiowa |
| OJT | On the Job Training |
| OP | Observation Post |
| OV-1 | Mohawk - Army Airplane |
| PCS | Permanent Change of Station |
| PFC | Private First Class |
| POW | Prisoner of War |
| PT | Physical Training |
| PTSD | Post-Traumatic Stress Disorder |
| PVT | Private |
| PX | Post Exchange |
| PZ | Pickup Zone |
| R&R | Rest and Recuperation |
| RC | Reserve Component |
| RON | Remain Overnight |
| S-1/G-1 | Personnel Staff Section |
| S-2/G-2 | Intelligence Staff Section |

Acronyms

| | |
|---|---|
| S-3/G-3 | Operations Staff Section |
| S-4/G-4 | Supply Staff Section |
| Stack arms | Stop Working |
| SF | Special Forces |
| SOG | Studies & Observations Group |
| Spec 4 | Specialist Four |
| STOL | Short Takeoff and Landing |
| TOC | Tactical Operations Center |
| TRADOC | Army Training and Doctrine Command |
| TRICAP | Triple Capability |
| U-1A | Otter – Army Airplane |
| U-6A | Beaver – Army Airplane |
| UCMJ | Uniformed Code of Military Justice |
| UH-IH | Utility Helicopter – Iroquois |
| USAR | United States Army Reserves |
| VC | Viet Cong |
| VFR | Visual Flight Rules |
| VNAF | Vietnam Air Force |
| XO | Executive Officer |

# Photo Gallery

**Ray's Parents, Ann & Raymond's Wedding Photo – 1936**

**Ray's Father – WWII**

**Grandma & Grandpa Dusek, Alan, & Raymond – 1941**

**Ray & Dad – 1942**

**Colonel Boland – Age 5**

**Ray "Dancing with the Stars," Age 6 – 1943**

**Brother Alan & Ray – 1943**

**Ray & Brother Alan – 1945**

**The Original Blues Brothers**

**Ray in his First year of ROTC – Age 13**

**(L-R): Master Sergeant Holsman, Eugene Croisant,**

**& Master Sergeant Mancini - 1954**

**Basic Training, Ft Leonard Wood – 1956**

**Ray & Jackie's Honeymoon – 1958**

**Part of a college trip with a stop in New York City**

**Commissioned as a Second Lieutenant at**

**Fort McCoy with his father Ray – 1959**

**8th Grade Class with Teacher Ray - 1961**

**B Battery 2/121 F.A. Marshfield, Wis. – 1961**

**Ray is circled**

**The Beavers Group – 1964**

**(L-R): George Volk, Shine Powell, Ray, & John Pfeiffer**

**Boland Boys Hanau, Germany – 1964**

**(L-R): Dave, Michael & Dan**

**Ray and L-19 Bird Dog in Germany – 1965**

**Ray with the Vietnamese neighborhood kids in Qui Nohn, Vietnam – 1966**

**The Otter aircraft at Pleiku, Vietnam –**

**December 1966**

**Mortar Attack Damage to Otter stationed at Pleiku –**

**January 1967**

**Captain Boland flying an Otter in Vietnam –**

**January, 1967**

**A remote village in Vietnam**

**Ray landed nearby and took a Jeep for a visit – 1967**

**Captain Boland flying a UH-6A Beaver from Fort Rucker to the Chicago Midway Airfield – 1967**

**Football Stars: Dan - 5th Grade & Dave - 3rd Grade – 1968**

**El Toro – 1970**

**Camp Eagle, Vietnam – 1970**

**Heroic Toros Pilots – 1970**

**(R - L): Joe Maxsom, Fred Cappo, Dick Femrite**

**Ray in the back seat of a Cobra Helicopter**

**Vietnam – 1970**

**Family Christmas after Second Tour in Vietnam – 1970**

**Family Photo at Fort Leavenworth – 1974**

**Ray on left, Assuming Command in Hawaii – 1978**

**Family Photo in Carlisle, Pa – 1981**

**Ray on left, End of Command Ceremony in Germany – 1987**

**Ray and Bob Hope – October 1990**

**Ray's Army Retirement**

**(L-R): General Hall, Ray & Jackie – 1991**

**The Toros Ride Again – 2023**

**(L-R): Back Row: Hackney, Fred Cappo, Dick Femrite, & Joe Maxsom; Front Row: Ray**

Made in the USA
Monee, IL
04 November 2026

67381617R00236